AF619318

Mathematics for the Curious Mind:

Stories From Everyday Challenges

Mathematics for the Curious Mind:

Stories From Everyday Challenges

Aalok Guptaa

Mathematics for the Curious Mind
Aalok Guptaa

Published in 2024

© Published by

Qurate Books Pvt. Ltd.
Goa 403523, India
www.quratebooks.com
Tel: 1800-210-6527, Email: info@quratebooks.com

ISBN: 978-93-58987-26-3

Book Creation Services

by

It's time you tell your story

www.bookmystorypublishing.com

A venture of

Contents

Preface

Let me guess, the last time you used math was when you were figuring out how much of your pay cheque could survive after a weekend splurge. Or maybe you've spent more time calculating how many hours of sleep you can squeeze in before that morning alarm, pretending it's the world's toughest math problem or the last time you used algebra was to figure out how much pizza you could afford after splitting the bill with friends. Don't worry, you're not alone. But here's the kicker–what if I told you that math isn't just for solving dinner bills or sleep schedules? It might just be the secret ingredient to making your career (and life) way more interesting than you'd ever imagine.

Let's be real–math has always had a reputation. For most of us, it's the subject that caused more sweaty palms than the school cafeteria's mystery lunch special. We've grown up believing math is that scary monster hiding under the bed, showing up only during exam season, armed with algebraic expressions and terrifying trigonometry.

But here's the plot twist: math isn't the villain in this story. In fact, it's been your quiet sidekick all along, helping you out in sneaky little ways. Think about it: when you're figuring out how much pizza to order for the group (because let's be honest, you don't want to be "that person" who ordered too little), or when you're trying to calculate just how many episodes of your favourite show you can binge-watch before bedtime and still function the next day—it's math, subtly working its magic.

And let's not forget, math has helped you survive life's greatest challenges—like navigating discounts during a shopping sale! You've probably become an expert in 20% off calculations faster than any math teacher could have predicted.

Still, math is often treated like that overly smart friend you feel awkward around—you know they're brilliant, but you're not quite sure how to talk to them without sounding, well, clueless. But here's the thing: that "smart friend" can help you in ways you never imagined. Want to crack the stock market? That's math. Dreaming of being the next tech billionaire? Yep, still math. Trying to figure out how many days until the weekend? Okay, that one's simple math, but still—math!

So, let's clear something up: math doesn't have to be that daunting character from a horror movie. In reality, it's more like a Swiss Army knife—ready to solve problems, open doors (literally and figuratively), and save the day in more ways than you've ever realized.

And trust me, after reading this book, you'll see math in a whole new light—not as a nemesis but as your personal secret weapon. Who knows? You might even find yourself *liking it*. Okay, let's not get ahead of ourselves, but hey, one step at a time, right?

I wasn't born a math whiz either. In fact, I was just like most people, convinced that math was the universe's way of torturing students. Back in the day, my friends and I had a shared understanding—math was that one subject we all feared. But things changed when my father, who saw something I didn't, encouraged me to take math seriously. Slowly, what started as fear turned into fascination, and over the years, I went from dreading it to loving it.

I remember my engineering days clearly. Math was the subject everyone feared, and somehow, I became the go-to tutor for my friends. It wasn't because I was a math prodigy—far from it! But once I started getting the hang of it, I found myself helping others survive their math-related nightmares. In return, my friends helped me through the subjects where I struggled. It was a fair trade—math for mechanics, equations for electronics.

The night before exams was always the same: caffeine-fuelled study sessions, desperate last-minute cramming, and a lot of bad jokes to ease the panic. We'd tell ourselves, "Just one more problem and then we're done," but somehow that never really happened. Instead, we'd pull all-nighters with instant noodles and hilarious debates over who could sneak snacks into the library undetected.

Those late-night sessions weren't just about math though—they were about facing the challenge together, finding humour in the madness, and realizing that math isn't some scary monster after all. And guess what? If I could survive those sleepless nights with a bunch of panicked engineers, you can tackle math too.

Today, math has become an integral part of my life—it's more than just a subject, it's a passion. Over time, I've come to realize that while many students study math simply to clear an entrance exam or qualify for professional courses, they often overlook the bigger picture. Math isn't just about passing exams—it's the foundation that can propel your career to new heights. Whether it's in cutting-edge fields like R&D, Artificial Intelligence, or Robotics, or even in something as simple and ubiquitous as a Google search, math is at the heart of it all. Every algorithm, every line of code—it's all based on mathematical principles.

Think about it. When you browse the internet, scroll through social media, or binge-watch your favourite series on a streaming platform, you're interacting with algorithms powered by math. From the way data is processed to how recommendations are generated, it's math working behind the scenes, crunching numbers faster than we ever could.

And let's not forget about innovation—self-driving cars, space exploration, even the latest advancements in medical technology are all possible because someone, somewhere, applied mathematical models to solve complex problems. Math is the common thread that connects these breakthroughs.

But it's not just the tech geniuses and researchers who rely on math. Even in everyday life—whether you're managing your finances, calculating interest rates, or planning your monthly budget—math is the silent hero helping you make informed decisions.

Take something as simple as the latest 'Mutual Fund Sahi Hai' campaign. It has gained tremendous popularity, and one of its key ideas is the Systematic Investment Plan (SIP), which is based on the principle of compounding. And what is compounding, at its core? It's math, plain and simple. Whether it's investing, calculating, or innovating, the principles of math are everywhere once you start paying attention.

At the end of the day, whether you're launching rockets or just trying to save for that dream vacation, math is always at play, shaping the world around you in ways you might not even realize. The best part? You don't need to be a genius to appreciate it. All you need is a little curiosity and an open mind—and who knows, you might just fall in love with math along the way.

From my personal experiences, I've come to understand that we deal with math every single day, often without even realizing it. Many of the world's greatest thinkers and business moguls, like Bill Gates, Mark Zuckerberg, and Elon Musk, share a deep appreciation for math. For them, math wasn't just a subject they studied—it was a tool that shaped their innovative journeys. Ultimately, coding, which has become so central to our daily lives, is nothing more than an extension of mathematics. And that's what I want to show through this book—math isn't just numbers, it's a gateway to endless possibilities in your career and beyond.

One of my primary goals in writing this book is to help people stop fearing math the way they do today. It's not a subject to be scared of—it's a tool, a way of thinking. Sure, there are formulas that can look intimidating, and many of us have been taught to memorize them without really understanding why they exist. But what if, instead of just memorizing, we tried to understand the concept behind the formula?

What if we stepped into the mind of the person who created it and asked ourselves, 'Why did they come up with this? How does it apply to the world around us?'

Once you grasp the reasoning and the real-world applications of these formulas, you might find yourself falling in love with mathematics. This book is my attempt to simplify math, to make it approachable and relatable. I want to show you how math isn't just a collection of abstract concepts but something that can truly impact your daily life, your career, and even the way you think. If you understand the concepts and their practical uses, I'm confident you'll see math in a completely new light—and maybe, just maybe, become a different person because of it.

There are plenty of books out there that attempt to simplify math or explain its concepts, but I've rarely come across one that truly bridges the gap between understanding math at a conceptual level and applying it in everyday life. That's exactly what I aim to do with this book. I want students—especially those who feel intimidated by math—to see how these concepts can be applied practically, making their lives easier and their careers brighter.

This book isn't just for students—it's for teachers, too. I hope to inspire educators to rethink how they teach math. If we shift from rote learning to teaching how math can be applied in the real world, students will not only understand it better, but they'll also start to enjoy it. And here's the thing—once students leave the classroom and step into the real world, math becomes one of the most powerful tools they can have in their arsenal. It makes them more employable, helps them express their ideas clearly, and can even drive innovation and success in fields like R&D.

My goal is to show that math isn't just a subject, but a lifelong skill. By mastering its concepts and applications, students can truly make an impact in their careers and the world around them.

One of my hopes for this book is that it becomes more than just a one-time read. I want readers to come back to it repeatedly, not because they missed something the first time, but because they've connected with it deeply. That connection only grows when you revisit ideas over time. As I've said before, this book is all about understanding the concept of math and how its application can simplify and improve your life.

There are two books that have had a similar impact on me, which I've read over and over. The first is Jim Collins' Built to Last, and the second is Fooled by Randomness by Nassim Taleb. Both books shifted my perspective in powerful ways. Taleb's book, for example, taught me that making decisions without being informed and analytical can lead to failure—even if, at first glance, things seem to be working out. True and lasting success requires a deeper understanding of the world around us, and for that, analytical thinking is essential.

And where does analytical thinking come from? One of the key tools for making thought-through, informed decisions is mathematics. It's not just about numbers and formulas—it's about developing a mindset that helps you think logically, assess risks, and make better decisions in both your personal and professional life. The kind of stopgap successes that some people rely on might seem like wins at the moment, but they won't build a sustainable path forward. A thoughtful, analytical approach is what leads to lasting success, and math is an essential part of that process.

So, let's make this journey enjoyable—set aside all your fears and misconceptions about math. Instead of seeing it as something intimidating, let's approach it as a tool that can completely transform your life. It worked for me, and there's no reason why it can't work for you, too.

Math has the power to unlock opportunities, reshape the way you think, and open doors you didn't even know existed. So, let's dive in together and begin this journey toward understanding how math can not only simplify your life but also make it richer and more fulfilling.

Finally, do you know why was the equal sign so humble?Because it knew it wasn't less than or greater than anyone else!

Hey!!, if by the end of this book you still aren't convinced that math can be fun, well, at least you'll have some great formulas to impress people at parties. Trust me, nothing says 'life of the party' quite like explaining the beauty of a quadratic equation over dessert! So, buckle up and let's begin on this mathematical adventure together—who knows, you might just fall in love with it along the way.

Remember, life is a delicate dance between two essential forces—emotions and logic. Emotions guide our passions, our connections, our dreams. But it's logic that turns those dreams into reality, giving us the tools to navigate life's challenges, make informed decisions, and create the future we want.

This book is all about the power of logic—specifically, the logic of mathematics. It's here to show you that math isn't just about solving equations or passing exams. It's about understanding the world around us. Math is the language of the universe, and once you understand its concepts and how to apply them, you unlock a whole new way of thinking that can truly transform your life.

Imagine being able to approach problems—both big and small—with clarity, precision, and confidence. Whether it's making financial decisions, designing innovative solutions, or even planning your next big adventure, math is the quiet force that will always be on your side. It's not just a subject; it's a superpower.

So, are you ready to rethink the way you see math? To discover its hidden potential in your life and career? Trust me, this journey is going to be a lot more exciting than you think.

Let's get started and unleash the power of math to change your world—one concept, one formula, one application at a time. You're about to embark on an adventure where logic reigns, and the possibilities are endless. Ready to dive in?

Foreword

I am an engineer by education and a data analytics consultant by profession. My journey has been shaped by curiosity and a desire to solve complex problems–a mindset I inherited from my father, AalokGuptaa. It is a privilege and a deeply emotional moment for me to write the foreword for his book, a project that reflects his passion, determination, and dedication to mathematics.

Professionally, I've had the privilege of working on ground breaking projects, including data analytics initiatives for logistics giants and prestigious educational institutions. My career began as a data engineer at TCS, and over time, I transitioned into consulting, helping organizations unlock the power of their data. My passion for problem-solving has led me to exciting milestones, such as presenting a paper on a fetal heart rate detection system at an international conference and securing a published patent for the project. Yet, all these achievements stem from one thing: the ability to understand, analyse, and make sense of patterns—a skill rooted in mathematics.

Growing up, I watched my father embody the principles he teaches. From his meticulousness to his relentless pursuit of knowledge, he has always been an inspiration. One of my earliest memories is seeing his engineering black book—a treasure of equations, diagrams, and notes. This notebook symbolized his deep connection to problem-solving, a trait that has influenced my own professional journey.

When my father approached me to write this foreword, I was both surprised and honoured. Having seen him transition from a successful career in the jewellery industry to building his dream of teaching mathematics, I knew this book was more than a guide—it was a culmination of his life's lessons, experiences, and relentless dedication. In 2020, during the global lockdown, I witnessed him sit for hours each day, poring over books, making notes, and refining his understanding of math to prepare for teaching. His ability to reinvent himself at 50, while balancing responsibilities, is nothing short of inspirational.

This book is not just about numbers or formulas—it's about making mathematics accessible and relatable to everyone, from students to young professionals in a storytelling form making it more interesting. Mathematics, as my father beautifully explains, is the invisible force shaping our world. Whether it's the algorithms behind your favourite apps, the statistical models predicting global trends, or the optimization techniques powering AI, math is everywhere.

Of all the chapters in this book, the one on Statistics resonates with me the most. Working in data analytics, I've seen first-hand how critical statistics is in shaping decisions across industries. Whether it's predicting trends, optimizing processes, or solving complex problems, statistics is at the core of understanding and leveraging data. Statistics is not just about numbers—it's about finding stories within those numbers. In my field, correlation, regression, and variance are tools that help uncover insights hidden in vast datasets.

For those picking up this book, I have one piece of advice: read with an open mind. This isn't a dry academic text; it's an invitation to see mathematics in a new light—as a tool that simplifies complexity, unlocks potential, and drives innovation. You'll find stories, analogies, and examples that make math engaging, relevant, and even fun.

To my father, I want to say this: thank you for showing me that reinvention is possible at any age, that passion fuels purpose, and that knowledge shared is the greatest legacy. Your journey is a testament to the transformative power of perseverance and vision. I know this book will inspire its readers as much as your journey has inspired me.

So, to the readers, dive in without hesitation. By the time you turn the last page, you'll not only have a better grasp of mathematics—you'll also uncover its deeper purpose. This book's mission is to profess mathematics as a way of life, to show us that it's not something to fear or run from, but rather a powerful tool to embrace. With its engaging approach, it helps us see math not as abstract formulas but as a guiding force that simplifies complexities, solves real-world problems, and inspires innovation. Through these pages, you'll gain not just knowledge, but a renewed sense of how mathematics shapes the world—and your unique place within it.

Tanishka Guptaa

My Journey:

From the Arid Sands of Rajasthan to the City of Dreams

Rajasthan in the 1970s was a land of vast deserts and unyielding conditions, where life was often a struggle between survival and scarcity. The state is and wasalways known for its vibrant culture, majestic forts, and royal heritage, but beneath the grandeur lay the harsh realities for many of its inhabitants. In the small villages, families faced daily challenges—limited resources, minimal access to education, and the constant battle to make ends meet. For many, the promise of a better life felt like a distant dream.

It was in this backdrop that I was born in 1970. Raised in a modest household, my early years were marked by the determination of a family striving to rise above their circumstances. My parents, like many others in rural Rajasthan, faced the harsh socio-economic conditions of the time. Jobs were scarce, money was tight, and every day was a test of resilience. My father, with his background in accounts, had a natural affinity for numbers—a trait I like to believe I inherited from him. He could make sense of figures and calculations with ease, and watching him work sparked my early interest in mathematics. But it wasn't just his knack for numbers that left an impression on me. My father was one of seven brothers, and despite their individual successes, they always remained closely knit.

Our family was built on the strong foundation of togetherness, rooted in the values of a joint family culture. Whether it was celebrating achievements or supporting each other through tough times, there was always a sense of unity and mutual care. These values weren't just words; they were lived every day, shaping the way I saw the world. Growing up in such an environment taught me the importance of collaboration, loyalty, and being there for one another—principles that have guided me throughout my life and career.

I am the eldest among my siblings, now with two brothers and a sister. But there's a chapter from my early years that stays close to my heart, one that profoundly shaped our family. I had an elder sister who, at the tender age of thirteen, passed away due to a rare illness. She developed a condition that led to a hole in her heart, and back then, medical science was not as advanced as it is today. Despite all our efforts and the love that surrounded her, we lost her. It was a devastating moment for our family, leaving a void that could never truly be filled.

Just six months before this tragedy, my mother had given birth to another baby girl, a ray of hope in our family. However, this younger sister was adopted by a close relative—a practice that was quite common in those days, a gesture of love and support within the extended family. While the adoption brought happiness and strengthened bonds between relatives, the loss of our elder sister shortly afterward added a complex layer of emotions to our lives.

The memory of my elder sister, her smile, and the brief time we had with her remains etched in my mind. Her passing is a reminder of the fragility of life, but also the resilience of family—a force that helps us move forward, together, even in the face of loss.

******

Growing up in such an environment taught me the importance of collaboration, loyalty, and being there for one another—principles that have guided me throughout my life and career.

******

In many ways, the structure of our family mirrored the logic and order found in mathematics: just as numbers and formulas must work together harmoniously to achieve the right solution, our family's strength came from our ability to stay connected and support each other. It's a lesson I carry with me, both in life and in the work I do today.

With little opportunity for advancement, we decided to make a bold move in search of a better future.

When I was just four years old, my family uprooted themselves from the familiar, serene life of Rajasthan and travelled to the bustling city of Bombay, as it was known then—the city of dreams, opportunities, and endless possibilities. Bombay was a place where fortunes were made, where every person who arrived carried the hope of building a better life. For us, too, that hope far outweighed the struggles that lay ahead.

The Bombay we arrived in was a complex, living organism—vibrant yet harsh. It was a city of contrasts. On one hand, there were towering colonial buildings, a relic of the British era, while on the other, there were crowded chawls, narrow lanes bustling with activity, and sprawling textile mills that were the backbone of Bombay's working class. The mills were where thousands of people toiled, from dawn till dusk, their livelihoods tied to the rhythm of the factory machines.

Bombay was also the land of the great mill strikeswhere workers fought for fair wages and better working conditions. These strikes would sometimes paralyze entire neighbourhoods, bringing both struggle and solidarity to the working class. The spirit of the people was resilient though—no matter how many obstacles came their way, they persevered. And this was the city we found ourselves in, with its unrelenting hustle and perpetual motion.

The social fabric of Bombay was diverse. People from all corners of India converged here—Marwaris, Gujaratis, Parsis, South Indians, and Maharashtrians, all blending into the city's melting pot. Each group brought their own culture, traditions, and hopes. But life in Bombay was never easy, especially for families like ours, coming from a humble background. The city demanded grit and determination. Jobs were hard to come by, and even harder to keep. The cost of living was rising steadily, with rent often being a significant burden.

For my family, settling in wasn't a smooth transition. We lived in a modest home, packed into small spaces that were common in those days. Every penny was accounted for, every meal a testament to my parents' hard work and sacrifice. Yet, despite the hardships, Bombay had its way of making you believe that anything was possible. It was a city where even the smallest opportunities could turn into something bigger—if you were willing to fight for it.

I remember my father telling me stories of how Bombay was evolving, how industries like textiles, manufacturing, and trade were booming. The docks, the factories, the stock exchange—everything felt alive with energy and ambition. But it was also a city of sharp contrasts—where the rich lived in grandeur, while the poor struggled to make ends meet. Amidst all of this, my father managed to carve out a space for us, relying on his knack for numbers and his accounts background. His passion for numbers became my inspiration, unknowingly setting the course of my life.

Bombay's streets, with their constant buzz of people, the clatter of local trains, and the ever-present smell of freshly brewed chai, became the backdrop of my childhood. It was here that my journey truly began, shaped by the relentless pace of the city and the values of perseverance and resilience passed down by my family. Bombay wasn't an easy place to settle in, but it was a city that rewarded hard work, and that's where my family found both challenge and opportunity.

******

In many ways, the structure of our family mirrored the logic and order found in mathematics: just as numbers and formulas must work together harmoniously to achieve the right solution, our family's strength came from our ability to stay connected and support each other.

******

After completing my schooling and earning a degree in Engineering with a specialization in Instrumentation, I took my first step into the professional world by joining a company called Inter Gold, a leading name in the jewellery manufacturing industry. I landed this opportunity through one of my uncles, who at the time worked for the real estate giant D B Realty.

In just six months at Inter Gold, I was immersed in the intricacies of the manufacturing process. The experience was like being handed the blueprint to running a business. I was exposed to every aspect–right from procurement, where we sourced the finest materials, to design and prototyping, where creativity met precision. I witnessed first-hand how operations were streamlined to ensure efficiency, how marketing crafted a brand's story, and how sales brought those stories to life in the marketplace.

It was a dynamic environment, and each department had something valuable to teach. However, amidst all the hustle, I found my true calling–processes and analysis. While most would be dazzled by the final product, for me, it was the underlying systems and operations that held a deeper fascination. I was intrigued by how each cog in the machinery, whether physical or metaphorical, worked in tandem to create something remarkable. Analysing data, optimizing processes, and ensuring smooth operations became my passion, and it was this foundational experience that set the course for the rest of my career.

Then came a pivotal moment in my career–one that I still regard as one of my most significant achievements. It was the implementation of SAP software in our company, a decision that transformed not only the way we operated but also my professional journey. SAP was introduced to the company by one of my colleague and me, and after careful deliberation, the management, including myself, embraced it wholeheartedly. It was a leap into uncharted territory, but the potential it held was undeniable.

I was entrusted with the responsibility of leading this monumental shift and was appointed as the Chief Technology Officer. The opportunity was exhilarating, but I quickly realized that it wasn't going to be an easy ride. The next two years were some of the most challenging yet rewarding in my career. Leadership was truly put to the test. It wasn't just about managing a new system; it was about managing people–navigating through resistance, guiding emotional responses, and creating attitudinal shifts in the organization. The stakes were high, and the pressure was immense.

Yet, through it all, I had an unwavering belief in the transformation that SAP would bring. I knew that if we could successfully integrate this system, it would revolutionize our processes, enhance efficiency, and sharpen operations. More than that, I could see the ripple effect it would have—customers would experience the benefits first-hand through more frequent product launches, innovative design options, and faster response times. It wasn't just about streamlining internal workflows; it was about elevating the entire value chain, right up to the customer.

As the implementation unfolded, every hurdle we faced felt like a test of endurance and resilience. There were days when leadership demanded not only strategic decisions but also empathy, as we had to win over teams and convince them that this change was for the greater good. But with each challenge came progress, and slowly, the vision started to materialize.

And just as I had envisioned, the transformation was profound. Customers were amazed and delighted by the newfound efficiency, innovation, and the variety of options we could offer them. In the late nineties, this kind of technological sophistication was unheard of in our industry. Our brand image and market positioning soared to new heights. SAP didn't just streamline our operations; it fundamentally changed how we were perceived by our customers and partners, strengthening relationships and boosting our bottom line.

This experience solidified my belief that true leadership lies in embracing change, trusting the process, and having the foresight to see how today's decisions can create tomorrow's successes. Looking back, it was more than just a technological upgrade—it was the moment where I realized the power of innovation, resilience, and forward-thinking in driving real, lasting change. It was a glorious association with Inter Gold for fifteen long years before I moved on.

My next professional chapter, though brief, was with the renowned Gitanjali Gems—a name synonymous with glamour and prestige in the jewellery industry. It was an exciting time, stepping into the role of Marketing Director, where I was at the forefront of driving the brand's image and market presence.

This role catapulted me into the world of high-end luxury, where I got the opportunity to witness and implement global standards. The exposure was immense, as I travelled extensively across continents, meeting industry leaders, understanding international market dynamics, and exploring new avenues for brand expansion.

Gitanjali was not just about crafting exquisite jewellery—it was about positioning the brand on a global stage. In my time there, I gained invaluable insights into global marketing trends, consumer behaviour, and the fine art of brand storytelling. It was a whirlwind of learning, travel, and high-level strategy, giving me a newfound appreciation for the intricate balance of luxury, culture, and commerce. Though my stint with Gitanjali was short, it left a lasting imprint on my career, broadening my perspective and refining my approach to business on a global scale.

Despite the glitz, glamour, and career-defining moments during my professional journey, one thing remained constant–my unwavering fondness for mathematics. You see, while most people traveling across the globe for work would dive headfirst into exploring new cities, soaking up the local culture, or ticking off tourist hotspots, I had my own unique (some might say eccentric) approach.

Instead of chasing famous landmarks or indulging in exotic cuisines, I found myself irresistibly drawn to the local communities. On weekends, while my colleagues might be sipping cocktails on a beach or taking selfies in front of iconic monuments, I was sitting at kitchen tables with children, teaching them math! Yes, I know, it doesn't sound like your typical globetrotting adventure, but for me, it was the best part of the journey.

In fact, my longest stint abroad, in Thailand, solidified this quirky passion of mine. For three years, I worked for a jewellery company there, and during my free time, instead of exploring temples or jungles, I was conducting impromptu math classes in neighbourhoods. I'd joke that I probably left behind more aspiring mathematicians than any travel photos! Even when I returned to India in 2018, my love for numbers never waned–it was still my way of connecting with people no matter where I went.

So, while others collected souvenirs, I collected memories of algebra equations solved and geometry problems cracked, leaving a trail of math enthusiasts in my wake. And you know what? It's the best souvenir I could have ever hoped for.

******

On weekends, while my colleagues might be sipping cocktails on a beach or taking selfies in front of iconic monuments, I was sitting at kitchen tables with children, teaching them math! Yes, I know, it doesn't sound like your typical globetrotting adventure, but for me, it was the best part of the journey.

******

After three transformative years in Thailand, I returned to India in 2018. Shortly after my arrival, a distant relative, Naresh Kejriwal, approached me with an intriguing proposition. Naresh had been running a successful jewellery business for a while, but he saw an opportunity to scale and innovate–and he believed my experience could help take things to the next level. Having known my background in manufacturing, processes, and especially my passion for precision (thank you, mathematics!), he was eager to team up.

This wasn't your typical jewellery store with glittering showrooms and walk-in customers. No, we operated from an office, meticulously crafting our designs, processes, and operations behind the scenes. It wasn't about quick sales or impulse buys; it was about relentless attention to detail, perfecting each product before it even reached a potential client's hands.

Our workdays were long and intense, focused on everything from design intricacies to refining processes that would make our products not just unique but exceptional. We didn't rely on foot traffic or flashy displays to sell our creations. Instead, we carefully curated our offerings and then invited select clients to view them, ensuring that each sale was as meaningful as the time and effort put into making the product.

It was a different approach, one rooted in craftsmanship, precision, and a deep understanding of our customers' needs. And though it required patience and hard work, it reinforced the values I had always held dear—attention to detail, passion for quality, and the belief that the process is just as important as the end product.

It was 2019, a year that started like any other but quickly became a defining moment in all our lives, thanks to the COVID-19 pandemic. The lockdowns and uncertainty brought many challenges, but for me, it also reignited an old passion—teaching mathematics, this time with a renewed sense of purpose.

As the world turned to online learning, I saw an opportunity to simplify math for students who were struggling with virtual classrooms and the subject's inherent complexities. I began recording math tutorials, one by one, with the goal of breaking down each concept into digestible, simple explanations. To date, I've recorded over ten thousand tutorials, each one focusing on the core concepts rather than just formulas. The objective was to take the fear out of mathematics and make it accessible to all.

This initiative gave birth to KkhyatisEdtech LLP, a venture I started with the mission of helping students from grade eight through grade twelve. It's not about cramming large classrooms; I focus on small batches of seven to eight students, offering them personalized attention and teaching. My approach is all about explaining the "how" and "why" of mathematics–simplifying formulas, clarifying concepts, and guiding students to see math as a tool rather than a hurdle.

Today, my sessions cater to a wide range of students–those preparing for board exams across CBSE, ICSE, SSC, and International boards like IGCSE, as well as students gearing up for competitive exams like MHT CET and other entrance exams such as IIT JEE and CAT. But it's more than just exam preparation. I see my role as a mentor, preparing students not just for their tests but for future careers in fields like Artificial Intelligence, Machine Learning, Robotics, Product Innovation, and Research.

*****

As the world turned to online learning, I saw an opportunity to simplify math for students who were struggling with virtual classrooms and the subject's inherent complexities. I began recording math tutorials, one by one, with the goal of breaking down each concept into digestible, simple explanations. To date, I've recorded over ten thousand tutorials, each one focusing on the core concepts rather than just formulas.

*****

While this endeavour has also helped me take care of my family financially, it was never driven by commercial intent. My focus has always been on mentoring and helping students unlock their potential in math. It's incredibly fulfilling to see these young minds get excited about a subject that used to scare them—and knowing that I'm helping to shape the future innovators of tomorrow.

My family has always been my pillar of strength, and I've drawn immense inspiration from my father, Lakhi Prasad Guptaa, a man whose love for numbers I seem to have inherited. His belief in the power of education and perseverance laid the foundation for my journey with mathematics.

I'm also fortunate to have an incredibly supportive partner in my wife, Sanjana Gupta. She's an innovative homemaker who's always stood by me through every twist and turn, cheering me on through each of my ventures. I still remember one of our school reunions, where she had the chance to meet my old friends. I thought they'd talk about my career achievements or the places I'd travelled, but to her surprise—and maybe mine too—they remembered me more for my math tutoring skills than anything else! There were endless stories about how I had helped them survive the dreaded math exams. Sanjana listened with a mix of amusement and pride, and by the end of the night, she was convinced that my passion for teaching math was more than just a hobby—it was my true calling. The seeds of KkhyatisEdtech were truly sown that evening.

I would like to express my heartfelt gratitude to my brother, Anup Gupta, whose unwavering encouragement and constant support have been the cornerstone of my journey. Whether it was through his wise words, his belief in my dreams, or his silent strength during challenging times, Anup has been my guiding light. His presence has not only inspired me to push boundaries but also reminded me of the value of perseverance and family. This book stands as a testament to the support of loved ones, and I dedicate this to him with immense love and appreciation. Thank you, Anup, for always believing in me when I needed it the most.

I'm blessed with two brilliant daughters who keep me grounded and inspired. Tanishka, my eldest, is an Electronics Engineer and is now pursuing her master's in Data Science. She's a numbers whiz in her own right, and I like to think some of that comes from my side of the family! My younger daughter, Khyati, is carving her path in Law, with dreams of becoming an Engineer Legal Advisor—a rare combination but one that perfectly suits her ambitions and intellect. Every time I see my daughters thriving, I'm reminded of how far the love of learning can take you.

So, when you see KkhyatisEdtech, know that it's not just a venture; it's a reflection of the values my family has instilled in me. It's the culmination of years of support, belief, and shared dreams. And while my friends may remember me for my math tricks, it's my family who has truly shaped the person I am today.

After diving into books like Jim Collins' *Built to Last* and Nassim Taleb's *Fooled by Randomness*, I found myself profoundly influenced by how great businesses—and, indeed, life itself—are built on time-tested principles. These books don't just discuss strategies; they reveal how small decisions, rooted in sound logic and understanding, create long-lasting success. For someone like me, who always sought clarity in numbers and patterns, these readings were nothing short of eye-opening. They sparked a thought: if I ever write a book, I want it to have that same powerful impact. I want to inspire readers to see the world differently and empower them with tools they can use daily.

With that, this book was born—not as a series of dry, formulaic lessons, but as a journey into the heart of mathematics. In the following chapters, I'll show you how math isn't just a school subject; it's a key to unlocking your potential. You'll discover where mathematics is hiding in your everyday life, waiting for you to apply it more effectively—whether in business, personal decision-making, or even creativity.

This isn't your typical math book. It's a roadmap to understanding how math can be a guiding force, and trust me, this journey will be unlike any trip you've taken before. So buckle up! You're about to embark on an adventure that will transform the way you think about numbers, logic, and life itself. Get ready for a ride full of introspection, inspiration, and excitement—because what lies ahead will challenge and amaze you in ways you never expected. Let's begin!

AalokGuptaa

https://kkhyatis.com/ , alokisgupta@gmail.com

Numbers in Action:

The Art of Practical Arithmetic

Have you ever found yourself mentally calculating how many coffees you can afford before your credit card bill hits, or trying to figure out whether that '70% off' sale is really as great as it sounds? If so, congratulations! You've already stepped into the world of arithmetic without even knowing it. Numbers aren't just for classrooms–they're everywhere, sneaking into your life like that last-minute online sale ofT-Shirt order on a Friday night. In this chapter, we're going to meet a few characters who, like you, are juggling their way through life with a little help from math. Don't worry, we're not here to memorize formulas–we're here to understand them in ways that will make your life (and maybe even your wallet) a lot simpler.

******

Rohan, a curious young student, had recently become captivated by an idea that seemed too good to be true. Everywhere he looked, from billboards on his way to school to pop-up ads on his favourite YouTube videos, he kept seeing the same promise: "Invest in a SIP today, and become a *crorepati!*"

Now, Rohan wasn't exactly sure what a "*crorepati*" really felt like, but he knew one thing–he wanted to be one. The ads made it sound so simple: just invest a small amount every month, and one day, you'd have a fortune big enough to make anyone's head spin. But Rohan had questions–lots of them. For instance, how could something as small as ₹10,00 a month turn into a crore? It sounded like magic. Naturally, he decided to ask his father for some answers.

His father, a busy professional in a pharma company, was used to fielding all sorts of questions from Rohan. But he wasn't quite expecting a deep dive into financial planning during dinner.

"Dad," Rohan began, "how can just ten thousand rupees a month make you a *crorepati*? Is it really magic?"

His father chuckled. "Magic? Well, if by magic you mean 'compound interest,' then yes, I guess you could say that," he replied, setting his plate aside to explain. "You see, Rohan, when you invest regularly in a SIP, your money doesn't just grow by what you put in. It grows on itself, thanks to something called compounding. It's a little like planting a tree–at first, it's just a sapling, but with time, it grows big and even bears fruit. And those fruits... well, they make more trees."

Rohan's eyes widened. "So, the money makes more money, which then makes more money?"

"Exactly!" his father said. "Think of it this way: if you invest ₹10,000 a month and keep it up for 25 years, the interest earned in one year doesn't just sit there–it gets added to your principal amount, and the next year, you earn interest on both. It keeps building up. That's why, even though you're investing small amounts, over a long time, it snowballs into a lot more."

Rohan's mind raced. He imagined planting a whole forest of money trees that kept multiplying, each giving a little more than the last. "So, SIPs aren't magic–they're like a really long science experiment?"

His father laughed. "In a way, yes! But a very rewarding one. You just have to start early and keep at it."

Armed with this newfound understanding, Rohan felt like he'd discovered a secret treasure map. All he needed to do was stick to the plan and let his money grow–and maybe, one day, he'd be that *crorepati* on the billboard.

Let me now demonstrate how this works practically:

"Alright, Rohan, let's take it up a notch" his father said. Imagine you invest 10,000 each month in a SIP with an expected annual return of 10%. Over 25 years, thanks to compound interest, this amount could grow significantly," his father began, grabbing a pen and paper.

“Here’s the formula we’ll use for the future value of a SIP:

$$FV = P \times \frac{((1+r)^n - 1)}{r}$$

Where:

- **FV** is the Future Value of your investment,

- **P** is the monthly investment (₹10,000 in this case),

- **r** is the monthly rate of return (annual rate of 10% divided by 12 months = 0.833% per month or 0.00833 in decimal),

- **n** is the total number of months (25 years X 12 = 300 months).”

Rohan’s father plugged in the numbers:

1. **P** = 10,000 (monthly investment)

2. **r** = 0.00833 (monthly rate of 10% annual return)

3. **n** = 300 months (25 years)

After calculating, he said, “This would give us a future value of approximately ₹1,32,20,000.”

Rohan's jaw dropped. "So, by putting in just ₹10,000 every month, I'd end up with over 1 crore rupees in 25 years?"

"Yes! That's the power of compounding," his father replied. "Of that total, ₹30,00,000 is what you've invested over time. The rest–over 1 crore–is what your investment earned through compounding!"

Rohan could hardly believe it. "So I just have to keep investing consistently, and compounding does the rest?"

"Exactly! Time and consistency are your best friends in building wealth," his father smiled. "The earlier you start, the more your money grows on itself, making you that *crorepati* you were dreaming of."

Graphically, this is how it looks:

The curve illustrates the power of compounding, especially in the later years (from the 150 months period onwards), where the growth accelerates significantly. This visual demonstrates how staying invested for the long term amplifies returns, as compounding has more time to work its magic.

While the growth chart of compound interest is impressive, the reality is that investment curves aren't always smooth. Markets have a natural rhythm—they rise and fall, sometimes sharply, and month-to-month returns can vary significantly. Unlike a steady, upward curve, real investments are often a mix of highs and lows.

For example, with a SIP of ₹10,000 per month, you might see some years where your returns soar, giving you a sense of rapid growth, only to be followed by months where values dip due to market corrections. This fluctuation can be unsettling, but it's precisely this market movement that makes compounding work in your favour over the long run. When prices dip, your fixed monthly investment buys more units, allowing you to accumulate assets at a lower price, which then grow when the market recovers.

In the long run, these fluctuations average out, resulting in what we call a **Compounded Annual Growth Rate (CAGR). CAGR** smooths out these short-term ups and downs, reflecting the average annual growth over time. This rate gives you a realistic view of your investment's growth and is often what investors use to gauge returns over extended periods.

So, while the journey may be bumpy, the destination is promising. By sticking with regular investments and allowing compounding to work through the ups and downs, you harness the long-term benefits of the market. The key is patience—stay invested, let compounding do its work, and remember that the magic lies in the long-term journey, not in the day-to-day fluctuations.

Rohan was amazed, glad he'd asked his father that one question. As his father explained, Rohan's mind was already buzzing with ideas—he wanted to try out the formula himself, testing different amounts, time horizons, and return rates to see just how far this "magic of compounding" could go.

And that's something you can do too! Play around with different amounts, experiment with various time horizons, and test out different rates of return. The key is to find an amount you can invest consistently. You could even start with as little as ₹500 per month and gradually increase it over time as your finances allow. The important thing isn't the starting amount—it's the commitment to staying invested and letting compounding work its magic over the long term.

So take your time, experiment, and remember: slow and steady wins the race. Compounding is a powerful ally, but only if you're patient enough to let it work on its own terms.

There are plenty of online tools and calculators available that you can explore to play with different investment scenarios. However, the real value lies in understanding the concept behind compounding and how it works over time. Once you grasp that, these tools become even more insightful.

And if you ever find yourself wanting a deeper explanation or a personalized approach, feel free to reach out to me. I'm always happy to dive into the details and help you see how math and compounding can work for your financial goals.

******

Here's a second story involving two Indian businessmen, Rajesh and Suresh, both working in the textile industry. This will allow us to explore the effects of credit on profits and the importance of understanding interest costs.

Rajesh and Suresh: A Tale of Credit and Cash Flow

Rajesh and Suresh had both been in the textile business for a few years, each running small shops in bustling markets. Though they sold similar products, their business practices were as different as night and day.

Rajesh's Credit Cycle Woes

Rajesh was known for his generous credit policy. He sourced his fabric and materials on credit terms from his suppliers, with payments due in 90 days. In turn, he offered the same 90-day credit terms to his customers, hoping that flexibility would build loyalty. But Rajesh hadn't factored one crucial thing into his business equation: the hidden costs of this extended credit.

When Rajesh calculated his profits, he looked only at the basic costs—the purchase price of the fabric and his overheads. He didn't account for the cost of delaying payments. With every passing month, he was, in effect, borrowing money to run his business, and each sale on credit tied up his cash flow further.

Rajesh didn't realize that with every month his customers delayed payment, he was incurring an implicit cost in the form of lost interest. If he had borrowed that amount from a bank at, say, a 12% annual interest rate, he would owe significant interest on the money tied up. His "profit" was slowly eroding, leaving him wondering why, despite good sales, his bank balance seemed stagnant.

Suresh's Cash-Only Strategy

Suresh, on the other hand, ran his textile business differently. He believed in a strict cash-only policy. He paid his suppliers in cash and didn't offer credit terms to his customers. Though some customers found this rigid, Suresh managed to attract enough business from those who preferred straightforward transactions.

By avoiding credit, Suresh kept his cash flow steady. He didn't need to borrow to cover operational costs or worry about delayed payments. His customers might not have liked it initially, but his pricing was slightly better than Rajesh's because he avoided the hidden interest costs that Rajesh incurred. Over time, his customers came to appreciate his transparent pricing.

The Impact of Interest Costs

To understand the impact of these two approaches, consider a scenario where Rajesh's business ties up ₹10,00,000 in credit sales, with a 90-day repayment period. If he incurs an implicit interest cost at a 12% annual rate, the interest cost over those three months would be:

$$\text{Interest Cost} = \text{Principal} \times \frac{\text{Annual Rate}}{4} = 10,00,000 \times 0.03 = 30,000$$

So, for every ₹10,00,000 tied up in credit, Rajesh effectively loses ₹30,000 in interest every quarter. Over a year, if his credit cycle remains the same, he could be losing around ₹1,20,000 without realizing it.

Meanwhile, Suresh, with his cash-based model, avoids these interest costs altogether, allowing him to reinvest his capital or pass the savings on to his customers.

Good cash flow management ensures that a business has enough liquidity to meet its obligations, grow, and handle unexpected expenses. Here's a closer look at the essentials of cash flow management:

1. Understanding Cash Flow Basics

- Cash Flow is the movement of money in and out of a business. *Positive cash flow* means more cash is coming in than going out, *while negative cash flow* means the business spends more than it brings in.

- A healthy cash flow is necessary to cover expenses like rent, salaries, utilities, raw materials, and debts.

2. Tracking Cash Flow

- Start with a Cash Flow Statement. This financial document breaks down your cash flow into three areas:

  - Operating Activities: Day-to-day activities, like sales and expenses.

  - Investing Activities: Cash from purchasing or selling assets.

  - Financing Activities: Cash from loans, repayments, or dividends.

- This statement helps you see where cash is coming from and where it's going, allowing you to monitor and adjust as needed.

3. Setting Payment Terms Wisely

- Offering credit terms to customers can attract more business, but it also ties up cash. Set clear credit terms (like 30, 60, or 90 days) and be aware of the impact delayed payments can have.

- If you're offering credit to customers, try to negotiate longer credit terms with suppliers, so you're not caught in a cash crunch.

4. Inventory Management

- Overstocking ties up cash and can lead to waste if products don't sell quickly. On the other hand, understocking can lead to lost sales. Use inventory management techniques, like Just-in-Time (JIT), to keep stock levels optimized.

5. Budgeting and Forecasting

- Cash Flow Forecasting involves predicting future cash flow to anticipate potential shortages. By forecasting, you can plan for slow months, make informed spending decisions, and know when it's feasible to reinvest or pay down debt.

- Budgeting ensures you have a spending plan that prioritizes essential expenses and aligns with your goals.

6. Handling Receivables and Payables

- To keep cash coming in, establish a system for managing Accounts Receivable (money owed to you). Send timely invoices, offer early payment discounts, and follow up on overdue accounts.

- For Accounts Payable (money you owe), try to take advantage of any payment terms offered by suppliers without jeopardizing relationships.

7. Financing Wisely

- Using external financing (like loans) can provide immediate cash flow but comes with costs. Evaluate options carefully and ensure that loan repayments align with cash inflows to avoid future cash flow issues.

- Consider revolving credit (like a line of credit) to handle short-term cash flow gaps.

8. Maintaining a Cash Reserve

- Having a cash reserve or emergency fund can help you handle unexpected expenses without impacting daily operations. Many businesses aim to keep at least 3-6 months' worth of expenses on hand.

Example:

Let's say Rajesh (from the previous story) has irregular cash flow because of extended credit terms. Here's what he could do to manage it:

- **Reduce Credit Terms:** Shift to a 30-day credit period to improve receivables' speed.

- **Negotiate with Suppliers:** Ask for a 60-day credit term from suppliers, reducing the pressure on cash.

- **Forecast Cash Flow:** Plan out monthly expenses and expected revenue to avoid surprises.

- **Build a Cash Buffer:** Save 10% of each sale as an emergency fund to cover months when cash flow is low.

Good cash flow management not only helps Rajesh stay operational but also provides room for growth, as he can invest surplus cash without risking his core operations.

By understanding and managing cash flow well, businesses can avoid debt traps, seize growth opportunities, and maintain stability, even in challenging times.

******

Let me now explain how even nature teaches us Mathematics.

The Fibonacci series is a sequence of numbers where each number is the sum of the two preceding ones, starting from 0 and 1. The series goes:

0,1,1,2,3,5,8,13,21,34,55,...

How the Fibonacci Sequence Works:

- The first two numbers are 0 and 1 by definition.

- Each subsequent number is the sum of the previous two numbers.

Mathematically, it's defined as:

Where:

- F(0)=0
- F(1)=1

Example:

- **Start:** 0,10, 10,1
- **Next**: 0+1=1
- **Then:** 1+1=2
- 1+2=3
- 2+3=5
- And so on...

Why is the Fibonacci Series Important?

The Fibonacci sequence appears in many natural phenomena and has significant applications in mathematics, art, and science:

- **Nature:** It's seen in the arrangement of leaves, flower petals, pinecones, and shells. For example, sunflower seeds form spirals based on Fibonacci numbers.

- **Mathematics:** The ratio of successive Fibonacci numbers approximates the **Golden Ratio** (approximately 1.618), which is considered aesthetically pleasing and appears in architecture, art, and design.

- **Computing and Algorithms:** The Fibonacci sequence is used in certain algorithms and in mathematical modelling, especially where recursive relationships are required.

In essence, the Fibonacci sequence is a simple yet powerful series that shows how patterns evolve through recursive growth, both in nature and in mathematics.

The Fibonacci sequence has fascinating applications in both day-to-day life and mathematics. Here's how it comes into play:

1. Financial Markets: Fibonacci in Technical Analysis

- In stock trading and financial analysis, Fibonacci levels are used to predict price movements. Fibonacci retracement levels, like 23.6%, 38.2%, 50%, 61.8%, and 100%, are derived from the Fibonacci sequence and Golden Ratio.

- These levels help traders identify potential support and resistance zones where prices might reverse. Though not full proof, they're widely used as a strategy in technical analysis.

2. Nature and Art: Patterns and Proportions

- **Natural Patterns:** Fibonacci numbers are observed in flower petals, pinecones, shells, and even DNA. The sequence helps explain growth patterns and spirals in nature, such as the arrangement of leaves on a stem or the spiral of sunflower seeds.

- **Art and Architecture:** Artists and architects often use the Golden Ratio (closely related to Fibonacci numbers) to create visually pleasing compositions. Paintings, sculptures, and structures that follow the Golden Ratio are thought to have an aesthetic appeal. The Parthenon and the Pyramids of Egypt are classic examples.

3. Mathematics and Algorithmic Design

- **Recursive Patterns:** Fibonacci is a great example of recursion in mathematics. Many problems in programming and algorithms, such as sorting and optimizing paths, are based on recursive techniques similar to the Fibonacci formula.

- **Algorithm Efficiency:** Fibonacci sequences are used in computer algorithms, especially for time complexity studies and in algorithm design to improve efficiency. It's commonly used to optimize data structures, dynamic programming, and search algorithms.

4. Daily Planning and Productivity: The Fibonacci Clock

- Some people use Fibonacci numbers in daily scheduling or productivity tasks. For example, using time blocks based on Fibonacci numbers (5 minutes, 8 minutes, 13 minutes, etc.) helps break up tasks into manageable, focused intervals, often called "Fibonacci timeboxing."

- This approach can make work more structured, allowing for natural breaks that keep the day productive without feeling overwhelming.

5. Gardening and Plant Arrangement

- **Planting Patterns:** Fibonacci patterns can guide gardeners in arranging plants for aesthetic appeal and efficient spacing. Following Fibonacci spirals can allow plants to get the most sunlight and water, making gardens both functional and beautiful.

- **Seed Spacing:** The spiral arrangement helps ensure that seeds and plants are spaced for optimal growth, especially in limited garden spaces.

6. Estimating in Project Management (Agile Development)

- In Agile project management, Fibonacci numbers are often used to estimate task complexity. Teams assign points in Fibonacci sequence increments (1, 2, 3, 5, 8, 13, etc.) based on a task's complexity and effort.

- The use of Fibonacci helps prevent "anchoring" and encourages a more balanced approach to project estimation by avoiding smaller incremental differences.

7. Educational Techniques: Teaching Recursive Thinking

- The Fibonacci sequence is a great example for teaching students about sequences, recursion, and exponential growth. It's often introduced early in math classes to develop an understanding of patterns, which can enhance logical and analytical thinking.

- Activities like arranging items in a Fibonacci sequence or creating simple programs to generate the sequence can help students develop a more hands-on understanding of recursion.

In essence, Fibonacci numbers aren't just theoretical; they provide valuable insight into patterns, structures, and behaviours in both natural and man-made systems, offering a way to optimize, predict, and create harmonious designs.

In each of these areas, the Fibonacci sequence helps create balance, predict outcomes, or optimize space, making it a surprisingly versatile tool beyond the world of math!

The Exponent of Arithmetic: Small Actions, Big Results

In both business and personal finance, arithmetic isn't just about basic addition or multiplication–it's a gateway to exponential growth. Exponential growth is what happens when small, regular efforts compound over time, leading to substantial results. It's a concept that often surprises people with just how powerful it can be.

Example: The Power of Doubling

Imagine if, on the first day of the month, you saved just 10. Each day after, you doubled the amount you saved from the previous day. Here's how it would look over a few days:

1. Day 1: ₹10
2. Day 2: ₹20
3. Day 3: ₹40

4. Day 4: ₹80

5. Day 5: ₹160

It's simple math, but by Day 10, you're saving over 5,000, and by the end of the month, the amount skyrockets into the millions! While doubling every day may not be realistic, it demonstrates the principle of exponential growth: the idea that consistent actions, even small ones, can accumulate into remarkable outcomes over time.

Exponential Growth in Business and Investments

In business, this concept is highly relevant. Just like a small, steady investment in a mutual fund can grow into a significant amount over years, investing in customer relationships, product improvements, or marketing efforts consistently over time can result in substantial long-term gains.

Exponential Growth Through Compounding

In finance, exponential growth is most commonly seen with compound interest. For example, we've already seen how with a modest investment of ₹10,000 each month in a savings plan that earns 10% per annum, you could see your initial contributions grow exponentially over a 20- or 30-year period. The magic of compounding kicks in as the interest builds on itself, allowing even modest investments to grow into considerable wealth.

******

Infusing Arithmetic with Indian Values

In India, we often grow up with lessons about patience, perseverance, and the value of small, consistent efforts—qualities that are as much a part of our upbringing as they are essential to mastering arithmetic. These values remind us that success, whether in personal growth, finance, or business, is less about quick wins and more about steady progress over time.

1. Patience and Consistency: The "Bachat" (Saving) Mindset

- Indian households often emphasize the importance of saving regularly, even if it's in small amounts. Whether it's putting aside a little money in a "gulak" (piggy bank) as children or learning from our elders about recurring deposits, we are taught early on that small, consistent efforts can lead to big results.

- The same is true in arithmetic and finance. Just like compounding interest, every small investment or saving effort adds up over time. By applying this "bachat" mindset to investments, savings, and even education, we embrace the wisdom of consistency and patience.

2. The Importance of Financial Discipline: "Saiyam" (Restraint, self-control, patience, endurance. refraining, abstinence with context to living within means)

- Traditionally, Indian families have encouraged a lifestyle of "saiyam," or moderation, prioritizing needs over wants. Arithmetic, especially when applied to budgeting, planning, and investing, becomes a tool that helps us live within our means while gradually building wealth.

- Whether it's setting aside money for children's education, a wedding, or retirement, financial discipline allows families to create a secure future. By calculating interest, forecasting savings, and managing cash flow, we ensure that these values are woven into the very fabric of our daily lives.

3. Preparing for the Future: "KalKeLiye" (For Tomorrow)

- Indian culture places a strong emphasis on preparing for the future. Parents and grandparents often make investments with the next generation in mind. From gold purchases to fixed deposits, the concept of "kalkeliye" teaches us to think long-term, laying the groundwork for stability and opportunity in the future.

- Arithmetic, particularly through concepts like compound interest, budgeting, and profit margins, helps us visualize and plan for these future needs. Just as our parents and grandparents invested for us, arithmetic empowers us to do the same for our children, ensuring that we leave behind a legacy of foresight and careful planning.

4. Value of Relationships and Trust: "Vishwas" (Trust) in Financial Deals

- In Indian culture, trust is paramount in business and finance. From informal loans among family members to credit offered within close-knit communities, many financial dealings are based on relationships rather than formal contracts. Arithmetic helps us manage these relationships carefully, tracking loans, calculating fair interest, and honouring commitments.

- This "vishwas" in relationships and financial dealings is a uniquely Indian value that's crucial to maintaining harmony in both personal and business matters. By applying arithmetic to ensure fairness and clarity, we honour this tradition of trust.

5. Embracing Frugality: "Kum Mein Zyada" (Getting More Out of Less)

- The concept of "kummeinzyada" or "getting more out of less" is deeply rooted in Indian culture. It's about making the most of resources, whether in the kitchen, at work, or in finances. Arithmetic helps us make wise decisions to stretch our resources, invest smartly, and maximize returns without unnecessary risk.

- This mindset reflects a commitment to resourcefulness and efficiency, values that can drive better financial planning, improved savings, and smarter investments.

The Takeaway: Arithmetic as a Path to Traditional Wisdom

For generations, Indian values have taught us the importance of small, thoughtful actions that build toward a bigger goal. Arithmetic is simply the tool that brings this wisdom to life. From saving small amounts each month to understanding how compounding works in our favour, arithmetic, combined with Indian values, empowers us to build a secure, prosperous future.

In this way, arithmetic isn't just numbers–it's a way of aligning with our roots, honouring the values passed down to us, and applying them to modern financial practices. Whether we're managing a business or planning for our family's future, these values are the guiding light that helps us use arithmetic in meaningful and impactful ways.

******

As we've seen, arithmetic and its principles are woven into the very fabric of our daily lives, from the simple act of budgeting to the magic of compound interest and the rhythm of natural patterns like the Fibonacci sequence. Understanding these concepts doesn't just make us better at numbers; it opens our eyes to a world of possibilities where we can make informed, confident decisions.

The beauty of math lies in its power to simplify complexities, helping us navigate everything from personal finances to future investments and beyond. Embrace these tools, play with the numbers, and let arithmetic become your ally on this journey. Who knows? The next time you see a number, you might just see a hidden opportunity waiting to unfold.

Let's move forward with curiosity and a readiness to explore even more of what math has to offer. This is just the beginning—so get ready, because the journey is about to get even more interesting!

Should you need further assistance in understanding any of these concepts, reach out to me.

Vectors and Matrices :

The Power Duo

Radhika and Ria were lounging on the comfy couch in Ria's house, sipping on their iced teas. The evening sun streamed through the window, casting a warm, golden glow across the room. Family photos adorned the walls, and the soft hum of the ceiling fan filled the air with a gentle breeze.

Radhika: "Ria, so that's so cool! You're studying engineering? I've always wondered how computers work and how websites come to life. Is it all magic or is there some real science behind it?"

Ria: "Well, it's a mix of both, actually! There's a lot of math involved, especially linear algebra. Concepts like vectors and matrices are fundamental to understanding how computers process information."

Radhika, intrigued: "Vectors and matrices? Sounds complicated!"

Ria, smiling: "Don't worry, it's not as scary as it sounds. Imagine a vector as an arrow pointing in a specific direction. It has a magnitude (length) and a direction. In computer graphics, vectors are used to represent points, lines, and surfaces. Matrices, on the other hand, are like grids of numbers. They're used for various operations, from solving systems of equations to transforming images."

Radhika: "And here I thought math was just something that gave me headaches!"

Ria: "Trust me, you're not alone. But remember, every time your GPS gets you out of a traffic jam, you can thank a matrix for that. It's like having a mathematical superhero in your pocket!"

Radhika, still curious: "So, how do these mathematical concepts relate to something like Google Search?"

Ria: "That's a great question! When you search for something on Google, the search engine uses complex algorithms to find the most relevant results. These algorithms often involve matrix operations. For example, the search engine might represent each webpage as a vector. By comparing the query vector (your search term) to the document vectors, the search engine can rank the results based on their similarity."

Just then, Rohan, Ria's father, joined the conversation.

Radhika, turning to Rohan: "Rohan *kaka,* Ria was just telling me about vectors and matrices. It sounds fascinating! Can you explain it in simpler terms?"

Rohan, chuckling: "Sure, Radhika. Think of vectors and matrices as tools that computers use to understand and manipulate information. They're like the building blocks of many complex systems. For instance, in machine learning, algorithms use matrices to analyze large datasets and make predictions. It's a powerful way to extract insights from data."

Radhika, amazed: "Wow, that's incredible! I never thought math could be so practical."

Rohan: "Exactly. Math is the language of the universe, and linear algebra is a particularly important dialect. It's the foundation for many technological advancements we see today."

Ria, her eyes wide with curiosity, turned to her father. "Dad, was Abhimanyu, in the epic Mahabharata, a victim of not knowing matrix applications? I mean, if he could have calculated the trajectory of the arrows and the formation of the Chakravyuha using matrices, maybe he could have escaped!"

Rohan chuckled, his eyes twinkling with amusement. "That's a very interesting thought, Ria. While matrices are a powerful tool, they are a product of modern mathematics. In ancient India, people relied on a different kind of knowledge, a deeper understanding of the universe and its laws. Abhimanyu's predicament was a result of his courage, his skills as a warrior, and the complex circumstances of the war. It was a test of his character and his abilities, not a failure of mathematical knowledge but yes it could have been the solution for him."

He paused, then added, "However, it's fascinating to think about how ancient wisdom and modern mathematics can complement each other. Perhaps, if we apply our understanding of the world and our knowledge of mathematics, we can solve the complex problems of our time."

Radhika, her eyes reflecting a mix of curiosity and concern, turned to Rohan. "With all these wars happening, especially the recent one in Israel, do you think vectors are being used to detect and intercept missiles? I mean, they can surely calculate the trajectory and pinpoint the target, right?"

Rohan nodded thoughtfully. "Absolutely, Radhika. Vectors are a fundamental tool in many modern technologies, including military applications. By analyzing the velocity, direction, and acceleration of a missile, engineers can predict its trajectory and potential impact point. This information can be used to deploy countermeasures or intercept the missile before it reaches its target."

He paused, then continued, "Similarly, in operations like the one that led to the elimination of Osama bin Laden, advanced surveillance and tracking technologies were employed. While the exact details of these operations are classified, it's reasonable to assume that mathematical techniques, including vector analysis, played a crucial role in locating and neutralizing the target."

Radhika, impressed, asked, “So, is it safe to say that vectors are the silent heroes behind many significant events in history?”

Rohan smiled. “In a way, yes. While they may not be as glamorous as the soldiers or the diplomats, mathematical concepts like vectors are the backbone of modern technology. They enable us to understand the world around us, solve complex problems, and make informed decisions. So, the next time you see a news report about a missile strike or a successful military operation, remember the role of mathematics in shaping our world.”

This is how google Image search, YouTube and Play find valuable content in milliseconds

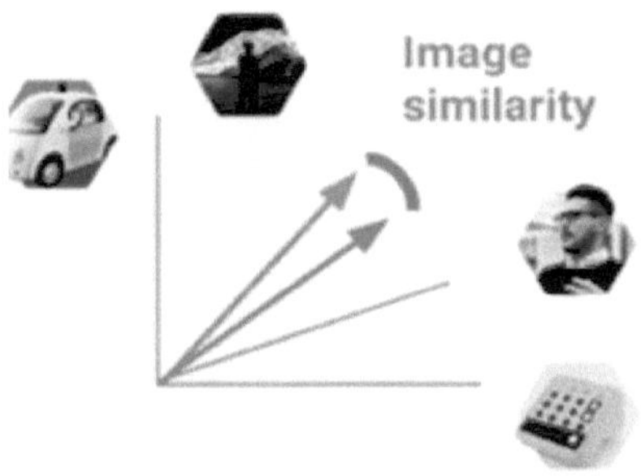

Source: https://cloud.google.com/blog/topics/developers-practitioners/find-anything-blazingly-fast-googles-vector-search-technology.

The room was abuzz with the intriguing conversation about vectors and matrices. Just as the discussion was reaching its peak, Anish walks in. A mischievous grin spread across Ria’s younger brother Anish’s face as he switched on the television, tuning into National Geographic.

“Hey! What do you think you’re doing?” Radhika exclaimed, reaching for the remote.

But before she could grab it, the TV screen filled with a stunning aerial shot of a majestic eagle soaring through the sky. “Woah, look at that!” she gasped, her attention instantly diverted.

Rohan, amused by the sudden shift in focus, chuckled. “Well, it seems nature itself is putting on a show. Let’s see what the experts have to say about it.”

As the documentary unfolded, the family was captivated by the intricate details of the eagle’s flight path, its hunting techniques, and its remarkable ability to navigate vast distances.

“Isn’t it amazing how nature has its own way of using vectors?” Ria pointed out. “The eagle’s flight path, its speed, and its direction can all be represented as vectors. It’s like a real-life application of the concepts we were just discussing.”

Rohan nodded in agreement. “Exactly. Nature is the ultimate mathematician. From the spiral of a seashell to the orbit of a planet, mathematical patterns are everywhere. It's fascinating to see how these patterns can be described and analyzed using the tools of mathematics."

The family continued to watch, their minds racing with new insights and connections. The once-abstract concepts of vectors and matrices were now taking on a new meaning, brought to life by the wonders of the natural world.

The evening was full of such discussions but the morning was no different.

The morning sun streamed through the window, casting a warm glow on Rohan as he sat engrossed in his laptop, a steaming cup of tea by his side. His eyes flickered between the television screen, displaying the fluctuating stock market indices, and the Excel sheet open on his laptop.

"Hey, "Radhika and Ria chimed in unison, their voices filled with curiosity. "That's matrices at work, isn't it? Crunching all that data, calculating those returns... it's impossible without them."

Rohan chuckled, a knowing smile playing on his lips. "You're absolutely right, girls. Matrices are the unsung heroes of the financial world. They help us analyze vast amounts of data, identify trends, and make informed investment decisions. Every time you see a stock market chart or a financial report, there's a good chance that matrices were involved in its creation."

He paused, taking a sip of his tea. "Think of it this way: each stock can be represented as a vector, with its price, volume, and other relevant factors as its components. By organizing these vectors into a matrix, we can perform complex calculations, such as correlation analysis and portfolio optimization."

Ria, her eyes wide with wonder, asked, "So, is it like a magic trick, Dad? How do matrices make it all so easy?"

Rohan chuckled. "Well, it's not exactly magic, but it's pretty close. Matrices are a powerful tool that allows us to manipulate and analyze data in efficient ways. By understanding the underlying mathematical principles, we can unlock valuable insights and make better decisions."

Rohan continued, "Exactly, Ria. Whether it's a seasoned stock market trader or a fund manager at a mutual fund company, their success often hinges on their ability to analyze data effectively. Matrices provide the framework for this analysis. They allow us to compare different stocks, assess market trends, and calculate risk factors.

Think about a business owner trying to maximize profits. They need to consider various factors like production costs, sales revenue, and market demand. These factors can be represented as vectors, and by performing matrix operations, the business owner can identify optimal strategies. For example, they might use matrices to determine the optimal production quantity or the best pricing strategy.

Ultimately, it's the human intellect that applies this knowledge and makes the final decisions. However, the power of matrices lies in their ability to provide a structured approach to problem-solving and decision-making. By leveraging the tools of linear algebra, we can gain a deeper understanding of complex systems and make more informed choices."

******

Welcome to the world of Vectors and Matrices.

Vectors and matrices, though abstract concepts, are the unsung heroes of our modern world. They underpin the intricate workings of computers, power the algorithms that drive our smartphones, and enable the stunning visuals in video games. These mathematical constructs, often visualized as arrays of numbers, are the building blocks of linear algebra, a branch of mathematics that deals with linear equations and their solutions. In this chapter, we delved into the fascinating world of vectors and matrices, exploring their properties, operations, and applications in various fields.

A vector is a mathematical object that possesses both magnitude (size) and direction. It can be visualized as a directed line segment, with the length representing the magnitude and the arrow indicating the direction. Vectors are fundamental in various fields of science and engineering, as they can be used to represent physical quantities like force, velocity, acceleration, and displacement.

Expanding on the Applications of Vectors

Physics and Engineering:

- **Mechanics:** Beyond representing forces, velocities, accelerations, and displacements, vectors are crucial in analyzing rotational motion, torque, and angular momentum. They also play a key role in understanding the principles of statics and dynamics.

- **Electromagnetism:** Vectors are used to describe electric and magnetic fields, as well as electromagnetic waves. They help visualize and quantify the interaction between charged particles and electromagnetic radiation.

- **Fluid Dynamics:** In addition to analyzing fluid flow and forces, vectors are used to study fluid properties like velocity, pressure, and vorticity. They are essential in simulating complex fluid behaviours, such as turbulence and fluid-structure interactions.

- **Computer Graphics:** Vectors are the foundation of 3D modelling and animation. They represent points, lines, and surfaces in space, enabling the creation of realistic and immersive virtual environments.

- **Robotics:** Vectors are used to control robot movements, path planning, and obstacle avoidance. They help robots perceive their environment, make decisions, and execute precise actions.

Other Applications:

- **Navigation:** Vectors are used in GPS systems to determine location, direction, and distance. They are also used in aircraft and ship navigation to calculate flight paths and optimize routes.

- **Meteorology:** Beyond analyzing wind patterns and weather forecasts, vectors are used to model atmospheric circulation, predict hurricanes and typhoons, and study climate change.

- **Game Development:** Vectors are used to simulate realistic physics, character movement, and camera angles in video games. They also play a crucial role in creating stunning visual effects and special effects.

- **Machine Learning:** Vectors are used to represent data points in high-dimensional spaces, enabling algorithms to learn patterns and make predictions. They are fundamental to various machine learning techniques, such as linear regression, support vector machines, and neural networks.

- **Data Science:** Vectors are used to visualize and interpret data, identify trends, and make data-driven decisions. They are used in data mining, clustering, and dimensionality reduction techniques.

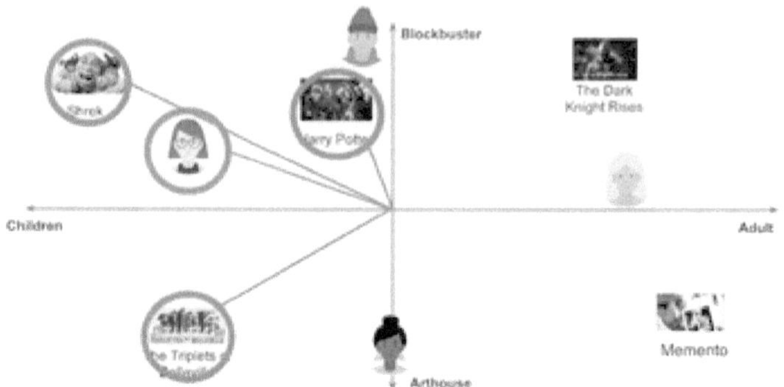

An example of a 2D embedding space for movie recommendation

Source: https://cloud.google.com/blog/topics/developers-practitioners/find-anything-blazingly-fast-googles-vector-search-technology

By understanding the versatile nature of vectors, we can appreciate their significance in shaping our modern world.

In essence, vectors provide a powerful tool for representing and manipulating quantities that have both magnitude and direction, making them indispensable in many areas of science, engineering, and technology.

In our increasingly interconnected world, where technology is reshaping industries and redefining human experiences, the power of vectors and matrices is undeniable. From the intricate algorithms that power artificial intelligence to the stunning visuals that bring video games to life, these mathematical concepts are the bedrock of innovation.

If you're passionate about delving deeper into fields like machine learning, artificial intelligence, robotics, gaming, or user interface design, understanding vectors and matrices is essential. These fundamental tools empower you to analyze complex data, build intelligent systems, and create immersive digital experiences.

If you're serious about pushing the boundaries of technology, understanding vectors and matrices is your golden ticket. These mathematical concepts are the building blocks of countless innovations, from self-driving cars to virtual reality experiences.

By mastering linear algebra, you'll be equipped to:

- **Revolutionize Healthcare:** Develop advanced medical imaging techniques, drug discovery algorithms, and personalized treatment plans.

- **Shape the Future of Finance:** Create sophisticated trading algorithms, risk assessment models, and fraud detection systems.

- **Advance Climate Science:** Analyze climate data, predict weather patterns, and develop sustainable energy solutions.

- **Pioneer Space Exploration:** Design spacecraft trajectories, optimize satellite orbits, and explore distant planets.

- **Enhance Human Experience:** Create immersive gaming experiences, design intuitive user interfaces, and develop intelligent virtual assistants.

The possibilities are endless. By understanding the fundamental principles of vectors and matrices, you'll be well-prepared to contribute to the next wave of technological advancements.

Ready to embark on this exciting journey? Let's connect and explore the limitless possibilities that lie ahead.

Mathematics of Motion:

The Wonders of Trigonometry

Yash and Rahul, two lifelong friends who had just passed their tenth-grade exams, were buzzing with excitement. Their trip to Dubai was not just a break but also a celebration–a gateway to a world filled with marvels that echoed their individual aspirations.

Yash, always captivated by architecture, dreamed of building skyscrapers that kissed the sky. Rahul, on the other hand, had his eyes on the stars, yearning to unravel the mysteries of space. Their friendship had always been rooted in shared curiosity, but they knew that the paths ahead would diverge as their passions led them in different directions.

The Flight to Dubai

As the plane roared to life and lifted off, the two friends settled into their seats, their conversation naturally drifting to Dubai's crown jewel—the Burj Khalifa.

"Imagine building something like that!" Yash said, his eyes gleaming. "The sheer height, the balance, the design—it's breathtaking!"

Rahul smirked. "Sure, it's tall, but I'm more interested in how satellites positioned above Earth use such structures for communication. It's all interconnected, you know."

Their discussion caught the attention of their fellow passenger, an elderly man with a warm smile sitting beside them.

"Fascinating, isn't it?" the man interjected. "The Burj Khalifa, the pyramids, satellites—they all owe their existence to mathematics, specifically, trigonometry."

"Trigonometry?" Yash and Rahul echoed, intrigued.

"Yes," he said, extending a hand. "I'm Professor Mahesh, a retired mathematician. Care to learn how these wonders connect?"

Exploring the Wonders of Geometry and Trigonometry

The professor began with the **Burj Khalifa.**

"Yash, to build such a marvel, engineers had to measure angles and distances to ensure stability. Trigonometry is used to calculate the forces acting on every part of the structure. The slope of the tower is carefully designed, using tangent ratios, to ensure it can withstand wind forces at its height of 828 meters."

He paused, letting the information sink in.

"And what about the pyramids?" Rahul asked, leaning in.

"The Great Pyramids," Prof. Mahesh said, "were built thousands of years ago, yet their construction also relied on basic trigonometric principles. The ancient Egyptians measured angles and used shadows–essentially a primitive form of tangent–to ensure precise alignments with the stars."

Space and Satellites

Rahul, whose heart was in the cosmos, couldn't resist steering the conversation toward his passion. "But how does this relate to satellites?"

The professor grinned. "Great question. Satellites rely on trigonometry for positioning. By triangulating signals from three satellites using sine and cosine functions, GPS systems can pinpoint locations on Earth.

Similarly, to calculate distances in space or angles for a rocket's trajectory, scientists depend on these principles."

Rahul's eyes widened. "That's incredible! So, whether it's Burj Khalifa or a satellite in orbit, trigonometry is the common thread?"

"Exactly," Prof. Mahesh replied. "It's the language that bridges architecture and astronomy, connecting your dreams, whether on Earth or among the stars."

New Horizons

As the plane began its descent into Dubai, Yash gazed out of the window, watching the city of skyscrapers come into view. Rahul, meanwhile, was lost in thought, imagining satellites silently orbiting high above.

"We might be walking different paths," Yash said, breaking the silence, "but we'll always be connected by something bigger."

"Like trigonometry," Rahul added with a grin.

Prof. Mahesh chuckled. "Indeed. And who knows? One day, Yash might design a space station for you, Rahul."

The trio laughed, their shared enthusiasm and curiosity weaving a bond that transcended their individual ambitions. As the wheels touched down in Dubai, they felt a newfound appreciation for the mathematical principles that had made their dreams—and the world around them—possible.

Their journey had just begun, not just as tourists in a dazzling city but as dreamers with their eyes on the horizon, ready to turn angles into achievements and ratios into reality.

******

Welcome to the world of Trigonometry.

Trigonometry is a branch of mathematics that studies the relationships between the angles and sides of triangles, particularly right triangles. At its core, it deals with the properties of angles and the functions that describe the ratios of their associated sides. These functions, such as **sine (sin), cosine (cos), and tangent (tan),** are fundamental tools used to solve problems involving measurements and calculations of distances, heights, and angles.

Morning came, bringing with it the scent of freshly brewed coffee and the chatter of excited tourists in the hotel's dining area. Yash and Rahul were among the first to arrive, their faces glowing with anticipation for the day ahead.

"Rahul, can you believe it? Today, we'll see the Burj Khalifa up close!" Yash exclaimed as they loaded their plates with food.

As they turned toward their table, they froze mid-step. Sitting there, with his signature warm smile and twinkling eyes, was none other than **Professor Mahesh.**

"Good morning, gentlemen!" the professor said, his voice as cheerful as the morning sun. "What are the odds of running into you two again?"

"Professor!" Yash exclaimed, setting down his plate in excitement. "What a coincidence! What brings you here?"

"A happy accident," Prof. Mahesh said with a chuckle. "I have a connecting flight to the U.S. tomorrow, so I thought I'd explore Dubai for a day. And now it seems I'll have the pleasure of your company for breakfast!"

Rahul grinned. "Looks like the universe really wants us to continue our conversation."

Ind ia's Gift to the World

As they sat down to eat, the professor leaned forward, an unmistakable glimmer of pride in his eyes. "Speaking of the universe, let me tell you about one of the greatest contributions India has made to the world of mathematics–something that ties directly to your interests."

Yash and Rahul exchanged curious glances, eager to hear more.

"Zero," Prof. Mahesh said, pausing for effect. "A number that might seem insignificant but holds the power to change everything."

"Zero?" Yash repeated. "You mean like... nothing?"

"Exactly! But nothing can mean everything," the professor said, leaning back with a knowing smile. "Long before the world even had a symbol for zero, Indian mathematicians were already using it to transform how we think about numbers, calculations, and the universe itself."

He continued, "It was **Aryabhata**, the great Indian mathematician and astronomer of the 5th century, who laid the groundwork. While his work didn't explicitly use the symbol for zero, he introduced positional notation, which is the very foundation of modern mathematics. Without it, we wouldn't have the number system we use today."

"Wait," Rahul interrupted, intrigued, "you're saying zero wasn't just invented to fill a gap in numbers? It was revolutionary?"

"Absolutely," Prof. Mahesh replied. "Think of zero as a placeholder–something that gives meaning to the numbers around it. For example, without zero, we couldn't differentiate between 1, 10, and 100. And it's not just about numbers. Zero represents balance, the midpoint between positive and negative, and even the concept of infinity in many ways."

Rahul, always the space enthusiast, leaned forward. "And Aryabhata–what else did he do?"

"Oh, much more!" the professor said. "Aryabhata calculated the value of **pi** to astonishing accuracy, described the Earth's rotation on its axis, and explained lunar eclipses–not through superstition, but science. His brilliance inspired countless others and set the stage for trigonometry, astronomy, and even space exploration."

The Burj Khalifa Connection

Yash, still pondering the conversation, asked, "So, Professor, does zero tie into things like the Burj Khalifa?"

The professor smiled, his gaze thoughtful. "Indeed. Without zero, there'd be no modern engineering, no skyscrapers, no computers to design them. Zero makes calculations possible—whether you're measuring the immense height of the Burj Khalifa or calculating the orbital trajectory of a satellite."

Rahul nodded. "And in space?"

"Everywhere," the professor said. "Zero is vital in calculating distances between planets, determining gravitational forces, and even programming the algorithms for space missions."

As breakfast came to an end, Yash and Rahul felt a newfound respect for the humble zero and the rich legacy of Aryabhata.

"Thank you, Professor," Yash said earnestly. "You've given us so much to think about."

"Indeed," Rahul added, "it feels like we're not just exploring Dubai. We're exploring history and the future."

The professor smiled, rising from the table. "And that, my dear friends, is the power of curiosity. It takes you places—whether to the stars, to the top of a skyscraper, or to a breakfast table in Dubai."

As the plates were cleared and coffee cups refilled, Yash couldn't help but voice a thought that had been nagging him.

"Professor," he began hesitantly, "why is there so much fear around trigonometry? I mean, so many of my classmates dreaded it, and honestly, even I didn't find it easy. What is it about trigonometry that makes it feel so... intimidating?"

Rahul nodded in agreement. "Yeah, it's like everyone panics the moment they hear about sine, cosine, or tangent. But when you explain it, it feels so logical."

The professor chuckled, leaning back in his chair as if savouring the question. "Ah, fear of trigonometry! It's a universal tale. But let me tell you a secret—it's not trigonometry itself that's the problem. It's how it's approached."

Both Yash and Rahul leaned in, intrigued.

The Professor's Explanation

"Think of mathematics as a vast forest," Prof. Mahesh began. "Each branch—trigonometry, algebra, geometry, calculus—represents a different tree. When students try to study trigonometry in isolation, it's like focusing on a single tree without seeing the entire forest. You lose perspective. The beauty of math lies in how these branches are interconnected."

"How do you mean?" Rahul asked, his curiosity piqued.

"Let me give you an example," the professor said. "Take trigonometry. On its own, it might seem like a collection of formulas—sin, cos, tan, and their inverses. But when you study it alongside vectors, you start to see its application in understanding direction and magnitude, whether it's calculating the force of wind on a skyscraper or plotting a rocket's trajectory."

He continued, "Then there's **matrices**—essential in computer graphics and simulations. They use trigonometric functions to rotate objects or manipulate 3D models. Without trigonometry, modern video games, virtual reality, and even animation wouldn't exist."

Rahul nodded. "That makes sense, but what about calculus? That's another subject people fear."

"Oh, calculus is like the lifeblood of modern science," the professor replied with a smile. "When you combine calculus with trigonometry, you unlock the ability to analyze motion, waves, and even orbits. Think of how we calculate the trajectory of a spacecraft—it's all about integrating the principles of trigonometric functions over time."

A Holistic View

"So, what you're saying," Yash said slowly, "is that trigonometry shouldn't be studied in isolation. It's part of a bigger picture."

"Exactly," the professor said, his eyes sparkling. "When you view mathematics holistically—connecting trigonometry with vectors, matrices, and calculus—you stop seeing it as a collection of daunting rules and start seeing it as a toolkit for solving real-world problems. Suddenly, it becomes not just manageable, but exciting."

He leaned forward. "Let me ask you this. Why do you think people are afraid of the dark?"

Rahul frowned. "Because they don't know what's in it?"

"Precisely," the professor said. "Fear comes from lack of understanding. Once you illuminate something, once you understand its purpose and connections, the fear disappears. Trigonometry is no different."

Perspective Shift

Yash and Rahul exchanged a glance, a silent understanding passing between them. What once felt like an intimidating subject now seemed like a stepping stone to understanding the world around them.

"So, Professor," Yash said with a grin, "if we want to truly master mathematics, we need to stop looking at it as separate subjects and start seeing it as a whole?"

"Absolutely," Prof. Mahesh said, his tone firm. "Math isn't a set of disconnected topics. It's a language—a way of interpreting and shaping the universe. And trigonometry is one of its most elegant chapters."

Rahul laughed. "I guess next time I see a sine or cosine, I'll remember it's part of something bigger."

"Do that," the professor said with a chuckle. "And the next time you're standing at the foot of the Burj Khalifa or gazing up at the stars, you'll see not just triangles and angles, but the very fabric of reality woven together by mathematics."

Their conversation shifted as they planned their day in Dubai, but the seeds of a new perspective had been sown. What once seemed like a mountain of daunting formulas now felt like a pathway to understanding the wonders of the world—and beyond.

******

Now let me get into more in detail on Trigonometry and it's applications:

How does trigonometry measure building heights?

Trigonometry measures building heights using the concept of right triangles. Here's how it works:

1. **Positioning:** Stand a known distance away from the building. For example, if you're 20 feet away and your height is 5 feet.

2. **Angle of Elevation:** Look up at the top of the building and measure the angle your line of sight makes with the horizontal. Let's say this angle is 30 degrees.

3. **Creating a Right Triangle:** This situation forms a right triangle where:

   o The height of the building is the opposite side.

- The distance from you to the building is the adjacent side.

4. **Using Trigonometric Ratios:** You can use the tangent function, which relates the opposite side (height of the building) to the adjacent side (distance from the building):

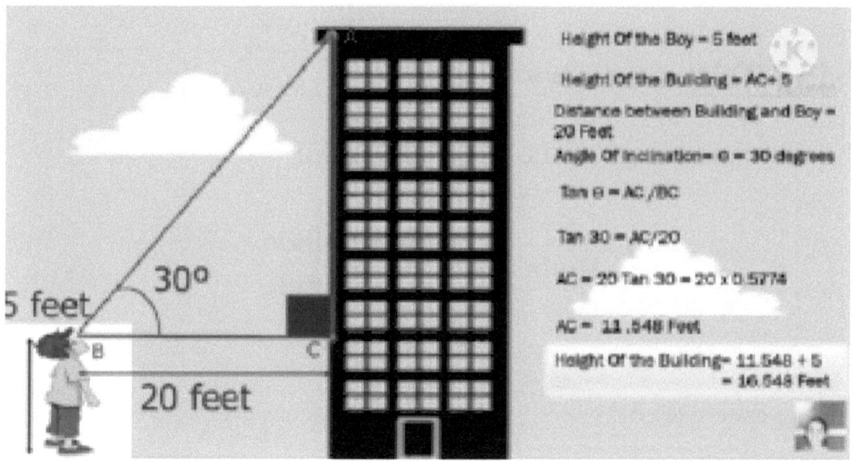

Source: https://www.youtube.com/watch?v=bGPIcQftgYI

This image demonstrates the use of **trigonometry** to calculate the height of a building using a simple observation setup. Here's how it works:

Scenario Explanation

1. **Observer's Height:** The person observing the building is 5 feet tall.

2. **Distance from the Building:** The observer stands 20 feet away from the building's base.

3. **Angle of Inclination:** The angle formed between the observer's line of sight to the top of the building and the horizontal ground is 30°.

Using Trigonometry

The calculation involves the tangent function:

$$\tan(\theta) = \frac{\text{Opposite Side}}{\text{Adjacent Side}}$$

Here:

- **Opposite side** (AC) = Height of the building above the observer's eye level.
- **Adjacent side** (BC) = Distance from the observer to the building (20 feet).

Step-by-Step Calculation:

1. **Apply the Tangent Function:**

$$\tan(30^\circ) = \frac{AC}{BC}$$

Substituting known values:

$$\tan(30^\circ) = \frac{AC}{20}$$

2. **Solve for AC:**

$$AC = 20 \times \tan(30^\circ)$$

From trigonometric tables: $\tan(30^\circ) = 0.5774$.

Therefore:

$$AC = 20 \times 0.5774 = 11.548 \text{ feet}$$

3. **Total Height of the Building**: Add the observer's height (5 feet) to AC:

$$\text{Total Height} = 11.548 + 5 = 16.548 \text{ feet}$$

Conclusion

Using the tangent function, the total height of the building is calculated as 16.548 feet. This example highlights the practical application of trigonometry in real-world scenarios, such as measuring inaccessible heights.

How is trigonometry used in criminology?

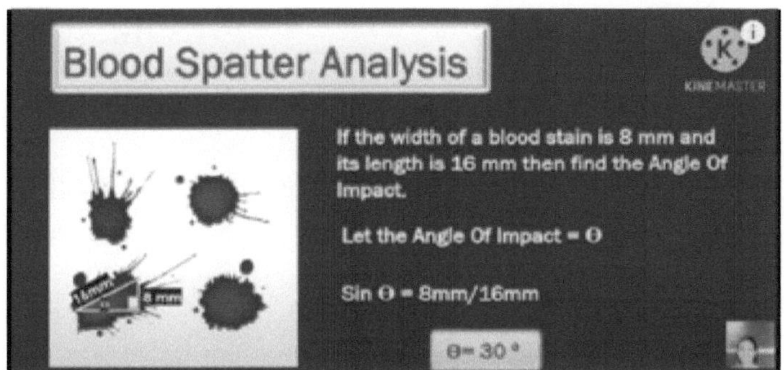

Source: https://www.youtube.com/watch?v=bGPIcQftgYI

This image demonstrates the application of trigonometry in blood spatter analysis, a critical aspect of forensic science used to reconstruct crime scenes. Here's a brief explanation:

Forensic experts analyze a blood stain to determine the angle of impact, which helps trace the origin of the blood. In this case:

- **Width of the blood stain:** 8 mm (opposite side of the triangle).
- **Length of the blood stain:** 16 mm (hypotenuse of the triangle).
- **Objective:** Find the angle of impact (θ).

Using Trigonometry

The sine function is applied here:

$$\sin(\theta) = \frac{\text{Opposite Side}}{\text{Hypotenuse}}$$

Substituting the given values:

$$\sin(\theta) = \frac{8}{16}$$

$$\sin(\theta) = 0.5$$

From trigonometric tables or a calculator, the angle whose sine is 0.5 is:

$$\theta = 30^\circ$$

Conclusion

The angle of impact is 30°, determined by analysing the dimensions of the blood stain using trigonometric principles. This technique helps forensic investigators understand the direction and origin of blood splatter, which can provide critical evidence in solving crimes.

In the world of crime investigation, popularized by fictional detectives like **Sherlock Holmes,** trigonometry plays a critical role in solving murder mysteries. Forensic experts apply principles like blood spatter analysis to reconstruct crime scenes. For instance, by calculating the **angle of impact** of blood stains, detectives can trace the origin of the blood back to a specific point in space. This helps determine the position of the victim or attacker at the time of the incident. Such precise mathematical analysis, combined with other forensic techniques, provides critical evidence to unravel the sequence of events, much like Holmes deducing the details of a crime with sharp observation and scientific reasoning. Trigonometry transforms abstract calculations into actionable insights, making the invisible visible in solving mysteries.

How is Trigonometry useful in navigation?

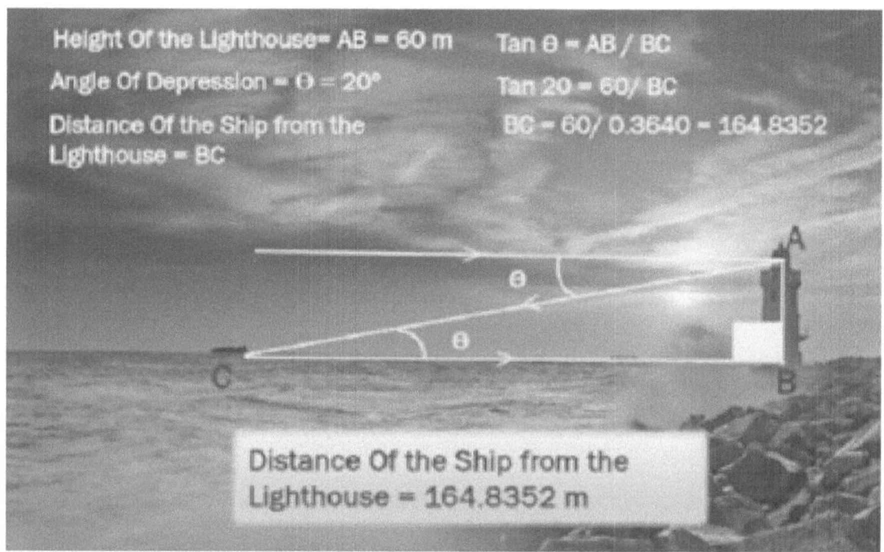

Source: https://www.youtube.com/watch?v=bGPIcQftgYI

This image illustrates the application of **trigonometry** to calculate the distance of a ship from a lighthouse using the **angle of depression.**

Scenario

- **Height of the lighthouse** (AB) = 60 meters.
- **Angle of depression** (θ) = 20°.
- **Objective**: Determine the distance of the ship (BC) from the base of the lighthouse.

Using Trigonometry

The tangent function is used:

$$\tan(\theta) = \frac{\text{Opposite Side}}{\text{Adjacent Side}}$$

Here:

- Opposite side = Height of the lighthouse (AB).
- Adjacent side = Distance of the ship from the lighthouse (BC).

Substitute the values:

$$\tan(20^\circ) = \frac{60}{BC}$$

Rearranging to solve for BC:

$$BC = \frac{60}{\tan(20^\circ)}$$

From trigonometric tables: $\tan(20^\circ) = 0.3640$.

Conclusion

The ship is approximately 164.84 meters away from the base of the lighthouse. This demonstrates how trigonometry is used in navigation to determine distances when direct measurement is not possible.

Usage in Space technology:

"Earth or Mars: The Next Address?"

Imagine a world where, instead of writing just your city, state, and country at the end of your address, you'd add **Earth** or **Mars**. Sounds like science fiction, right? Yet, it's a vision that **Elon Musk** is turning into reality. With his ambitious **Mars colonization program,** Musk is working tirelessly to make Mars not just a destination, but humanity's **second home.** He envisions a time when families will pack their belongings, say their goodbyes to Earth, and head off to Mars to build a new life under its dusty red skies.

But how is all of this possible? How do we study an alien planet, plan colonies, and transport people across millions of kilometers? The answer lies in **satellites, data,** and the incredible power of **mathematics**–particularly **trigonometry, vectors**, and **matrices.**

Satellites: The Eyes in the Sky

Every day, fleets of satellites launched by companies like SpaceX orbit Earth and venture further into space. These satellites are our explorers, silently gathering critical data about Mars, its atmosphere, and the surrounding universe. As you read this, they're sending streams of information back to Earth—distances, speeds, angles, trajectories, and even 3D maps of the Martian surface.

Without satellites, our dream of reaching Mars would remain just that—a dream. But how do we decode this data, make sense of it, and use it to make life-changing decisions?

Trigonometry: Unlocking the Universe

Here's where **trigonometry** comes in. Imagine this: a satellite is orbiting Mars, capturing images and measuring angles to calculate distances between craters, mountains, and valleys. Using these measurements, scientists create precise maps of the Martian surface, determining the safest landing zones for spacecraft or potential sites for human settlements.

For example:

- **Angles of elevation and depression** are used to calculate the height of Martian mountains and the depth of its craters.

- **Sine, cosine, and tangent functions** help scientists measure the distance between two locations without ever physically being there.

This isn't just math—it's the blueprint for survival on another planet.

Vectors: Guiding the Path to Mars

When rockets blast off from Earth, they don't travel in a straight line. Instead, their path is carefully plotted using vectors, which account for speed, direction, and gravitational forces from celestial bodies like the Moon and Sun.

- A vector points the spacecraft toward Mars while considering how Earth and Mars are constantly moving in their orbits. It's like aiming for a moving target while you're moving yourself–a task only possible with advanced vector calculations.

- Once on Mars, vectors guide rovers **like Perseverance** as they explore the planet. The direction and magnitude of their movement are controlled using vector principles.

Matrices: The Key to Understanding Data

Every image or piece of information sent back by satellites is made up of massive amounts of data. To process this, scientists use **matrices**– grids of numbers that allow them to organize, analyze, and transform this data into meaningful insights.

- For example, a satellite image of Mars is essentially a matrix of pixels. Using matrix transformations, scientists can enhance the image, identify patterns, and even simulate what the surface might look like in real life.

- Matrices also play a role in simulating potential environments on Mars, helping engineers design habitats that can withstand the planet's harsh conditions.

The Bigger Picture

As satellites continue their relentless mission to gather information, Musk's engineers and scientists back on Earth use these mathematical tools to bring his vision closer to reality. **Trigonometry** helps map the unknown, vectors ensure safe travel, and matrices turn raw data into actionable knowledge.

A Future Written in the Stars

Now imagine this: Years from now, a child born on Mars grows up to study the **Pythagorean theorem** and learns how it helped their ancestors measure the distance to the stars. That child writes their address on a piece of paper, and at the end of it, they proudly write **Mars.**

What once seemed crazy–a second home for humanity–is becoming real, driven by the boundless possibilities of mathematics and the unyielding human spirit to explore, discover, and thrive. **The universe is no longer out of reach—it's just another chapter waiting to be written.**

******

Trigonometry is not just a subject confined to textbooks; it's a universal language that connects the past, present, and future. From the ancient pyramids of Egypt to the towering Burj Khalifa, from navigating ships across oceans to calculating spacecraft trajectories to Mars, trigonometry is the silent architect behind humanity's greatest achievements.

It empowers us to measure the immeasurable, to see the invisible, and to unlock mysteries that were once thought to be beyond our reach. Whether it's reconstructing a crime scene, designing the next generation of skyscrapers, or charting a course to the stars, trigonometry proves time and again that it is the foundation of innovation.

As we dream of a future where Mars becomes an alternate home, trigonometry will continue to guide us—bridging the gap between imagination and reality. It reminds us that every angle, every ratio, and every calculation has the potential to shape the world, and even the universe, in ways we never thought possible.

So the next time you encounter a sine, cosine, or tangent, don't see them as mere formulas. See them as keys—keys to understanding the wonders of the world, to solving the greatest challenges, and to exploring realms yet unknown. Trigonometry is not just math; it's the art of turning dreams into achievements, one angle at a time.

We live in extraordinary **times** where the boundaries of human imagination and technological innovation are being pushed farther than ever before. The journey from measuring simple dimensions—length, breadth, and height—to incorporating the abstract concept of time, has now brought us to the brink of a new frontier: the **metaverse.**

But what does this evolution mean, and how is it shaping the way we interact with the world around us?

Now, as technology advances, we're stepping into the **metaverse**—a virtual, immersive, and interactive digital world that exists parallel to our physical reality. While the metaverse is not a literal fifth dimension, it represents a new way of experiencing reality that transcends traditional boundaries of space and time.

As humanity continues to reach for the stars, trigonometry will light the way, proving that the universe itself is a canvas—and mathematics is the tool to paint it.

Calculus:

The mathematics of change

Raj and Azhar walked out of the theatre, their eyes wide with awe after watching the sci-fi classic *Terminator*. The movie's thrilling depiction of killer robots, time travel, and futuristic technology had left them buzzing with excitement.

"Man, can you imagine robots actually predicting our every move?" Azhar exclaimed. "It's like they're calculating changes in real time to outsmart us!"

Raj chuckled. "It's crazy, but you know what? That's not just science fiction. What if I told you there's actual math behind all that?"

Azhar stopped mid-step, turning to his friend. "Math? Come on, Raj. Don't ruin the fun with numbers."

"No, seriously!" Raj said, his grin widening. "It's all about calculus. You know, derivatives, rates of change... it's how those robots in the movie could predict movements and react faster than humans."

"Wait a minute," Azhar said, his curiosity piqued. "Are you saying math like that can make robots *think*?"

"Not just think," Raj replied, with a spark of enthusiasm. "It's the same math that helps us send rockets into space, forecast weather, and even figure out how diseases spread. Calculus is the secret to understanding change–just like those robots did."

Azhar's eyes lit up. "So, basically, Calculus isn't just about solving equations in school. It's like... the code that runs the universe?"

"Exactly!" Raj said. "And the coolest part? It's not sci-fi. It's real."

As they continued walking, Azhar couldn't help but see the world differently. What once seemed like a dull math class now felt like the gateway to controlling time, space, and maybe even building a robot army of his own.

After finishing their coffee, Raj and Azhar parted ways, but Azhar couldn't shake the conversation about Calculus from his mind. As he walked home, snippets of their talk about robots, rockets, and predicting change kept replaying in his head.

That night, instead of reaching for his gaming console or scrolling endlessly on his phone, Azhar did something unexpected—he opened his laptop and typed "What is Calculus?" into the search bar.

At first, he was overwhelmed by the technical jargon. Limits, derivatives, integrals—it all sounded daunting, like the stuff he had skimmed through in school just to pass exams. But this time, Azhar wasn't studying for a test. He was driven by pure curiosity, and that made all the difference.

He stumbled upon a video explaining how derivatives are used to calculate the speed of a rocket at a specific moment or predict the trajectory of a basketball shot. Another link showed how integrals help architects calculate the exact amount of material needed to construct curved structures like domes and bridges. Slowly, the pieces of the puzzle began to fit together.

Azhar's mind raced with possibilities. "So this is how they build robots that can move like humans," he thought, recalling scenes from *Terminator*. "And it's how scientists figure out the best way to send satellites into orbit or predict tomorrow's weather."

The more he read, the more captivated he became. What once seemed like boring formulas on a classroom board now felt like the key to unlocking the secrets of the universe. For the first time, Azhar saw Calculus not as a subject to be feared, but as a powerful tool—a language that explained how the world worked.

That night, Azhar fell asleep with a notebook by his side, scribbled with newfound insights. His curiosity had been sparked, and he knew this was just the beginning of a deeper journey into understanding the math of change. Tomorrow, he decided, he'd ask Raj to explain more about this fascinating world of Calculus.

******

India's role in the development of Calculus dates back centuries, with mathematicians like Madhava of Sangamagrama laying the groundwork. Madhava, often called the "Father of Calculus," developed early concepts of infinite series to approximate values for trigonometric functions like sine and cosine. His pioneering work predated the contributions of Newton and Leibniz in Europe. Other Indian mathematicians, like Aryabhata, introduced ideas of limits and approximation, which became critical in modern Calculus. This heritage underscores India's profound impact on the mathematical tools we use today.

Predicting with Utmost Accuracy

Calculus enables precise predictions by analysing and quantifying change. Through derivatives, we can calculate instantaneous rates (e.g., speed at a specific moment), and with integrals, we measure accumulation (e.g., total distance travelled). For example:

- In finance, derivatives help predict stock price trends by analysing minute-by-minute fluctuations.

- In engineering, integrals allow for precise calculations of material needs for complex structures.

By continuously refining data through mathematical models, Calculus ensures predictions are not just estimates but highly accurate forecasts.

Calculus in Today's Dynamic World

In an ever-changing world, Calculus serves as a powerful tool for predictability, especially when visualized through graphs. By plotting data points:

- Businesses can forecast sales trends, identify optimal pricing strategies, and minimize costs.
- Daily Life Applications include analysing fuel consumption during a trip or predicting electricity usage based on past patterns.

Graphical representations, combined with Calculus, allow us to identify maximums, minimums, and inflection points, providing a deeper understanding of patterns. Whether it's optimizing a production line or determining the best time to invest, Calculus transforms raw data into actionable insights, making it indispensable in today's dynamic and data-driven world.

Here's a simple example of a manufacturer wanting to maximise volume and profits in a packing material.

Scenario

You want to design a rectangular box with:

- Open or closed top.
- Fixed material for the surface area.
- The goal is to maximize the volume.

Step 1: Define the Variables

1. Let the dimensions of the box be:

- o Length = l
- o Width = w
- o Height = h

2. Volume of the box:

V = l X w X h

Step 2: Express Volume in Terms of Two Variables

Using the constraint A - fixed area, solve for one dimension (eg. H) in terms of l and w.

For an open box:

$$h = \frac{A - lw}{2(l + w)}$$

Substitute h into the volume equation

$$V = l \cdot w \cdot \frac{A - lw}{2(l + w)}$$

Step 3: Differentiate to Find Maximum Volume

To maximize V, calculate its partial derivatives with respect to l and w, and set them equal to zero:

$$\frac{\partial V}{\partial l} = 0, \quad \frac{\partial V}{\partial w} = 0$$

Step 4: Verify the Maximum

Use the second derivative test or analyse the boundary conditions to ensure the critical points yield a maximum volume.

Example

Problem: Maximize the volume of an open box with a fixed surface area A = 120 cm2

1. **Surface Area Equation:**

$$120 = lw + 2lh + 2wh$$

2. **Volume:**

$$V = l \cdot w \cdot h$$

3. Substitute h in terms of l and w:

$$h = \frac{120 - lw}{2(l + w)}$$

4. Substitute h into V:

$$V = l \cdot w \cdot \frac{120 - lw}{2(l + w)}$$

Here's a video explaining with actual numbers to demonstrate: https://www.youtube.com/watch?v=DTEDY95V9Dc

Calculus is a branch of mathematics that studies how things change and accumulate. It provides tools to analyse motion, growth, and other dynamic phenomena, making it a cornerstone of modern science, engineering, and technology.

Core Components of Calculus

1. Limits:

- The foundation of Calculus.
- Describes the behaviour of a function as it approaches a specific value.
- Key to defining derivatives and integrals.

2. Derivatives:

- Concerned with the rate of change.
- Example: Speed is the derivative of distance with respect to time.
- Applications: Motion analysis, optimizing functions, understanding slopes.

3. Integrals:

- Deals with accumulation or the total effect of changes.
- Example: Area under a curve represents the integral of a function.

- Applications: Calculating areas, volumes, and total quantities.

4. Differential Equations:

- Equations that involve derivatives.
- Used to model complex systems like population growth, weather patterns, or electrical circuits.

Branches of Calculus

1. Differential Calculus:

- Focuses on finding derivatives.
- Studies rates of change, slopes of curves, and optimization problems.

2. Integral Calculus:

- Focuses on finding integrals.
- Explores accumulation of quantities and areas under curves.

3. Multivariable Calculus:

- Extends calculus to functions of more than one variable.
- Applications: Fluid dynamics, 3D modelling, and electromagnetism.

4. Differential Equations:

- Combines derivatives with equations to describe how quantities evolve.
- Applications: Engineering, physics, biology.

Significance of Calculus

- Enables us to model and predict real-world phenomena with precision.
- Foundational for fields like physics, engineering, economics, medicine, and data science.
- Tools like derivatives and integrals allow for analysing both small-scale and large-scale changes.

In essence, Calculus provides a framework to understand and describe the ever-changing world around us.

******

"The Calculus Chronicles: When Math Meets Reality"

Azhar couldn't wait to meet Raj the next day. His night of research had opened up a whole new world, and he had a thousand questions buzzing in his mind. As they met at their usual café, Azhar, brimming with excitement, barely waited for Raj to take a sip of his coffee before diving in.

"Raj, I spent the whole night reading about Calculus. You didn't tell me it's basically the magic wand behind everything cool in science!" Azhar exclaimed, pulling out a notepad filled with doodles of graphs, formulas, and a poorly drawn robot.

Raj chuckled. "Magic wand, huh? I guess you're starting to see what I was talking about yesterday."

"Starting to see? I'm blown away! Do you know they use Calculus to predict global warming, build robots, and even land rockets? I feel like I just uncovered the cheat code to the universe!" Azhar said, waving his notepad dramatically.

Robots, Global Warming, and the Math That Drives It All

Raj leaned back, smiling. "Okay, okay, slow down. Let's start with something fun–robots. You know those epic robots in *Terminator*? The way they predict movements, avoid obstacles, and even outsmart humans? That's Calculus in action."

Azhar raised an eyebrow. "How? I mean, they're robots. They just... do stuff."

"It's not that simple," Raj explained. "For example, when a robot's moving toward an object, it calculates its position, speed, and even changes in movement over time. This involves derivatives to figure out how fast something is approaching and integrals to calculate the path it needs to take to reach or avoid the object."

"And what about when the object's moving too? Like in that scene where the robot dodged a missile mid-air!" Azhar asked, miming a robot dodging an imaginary projectile.

"That's where trigonometry comes in," Raj said. "The robot needs to calculate angles and distances in real-time. Add in statistics, and it can predict where the missile will be next, based on its current trajectory."

Azhar shook his head, impressed. "Man, robots are basically nerds with perfect math skills."

Predicting the Future: Global Warming and Calculus

"Speaking of predicting," Raj continued, "let's talk about something that affects all of us–**global warming**."

Azhar's face turned serious. "Yeah, I read about that. Calculus helps forecast climate change, right?"

"Exactly. Scientists collect data on things like temperature, CO_2 levels, and ice melt over decades. Using statistics, they organize all this data into patterns. Then, with Calculus, they figure out rates of change–how fast the ice is melting, how much CO_2 levels are rising, or how sea levels will increase in the next 50 years."

"So, like, they're using derivatives to see how bad things are getting and integrals to figure out the total impact?" Azhar asked, catching on quickly.

"Bingo!" Raj said, giving Azhar a high five. "And the coolest part? This helps humanity plan ahead. Without those predictions, we wouldn't know how urgent it is to switch to renewable energy or prepare for rising sea levels."

Azhar leaned forward, looking thoughtful. "So, in a way, Calculus isn't just math–it's like a superhero, saving the planet one equation at a time."

Raj laughed. "I guess you could say that. Though it's more like a behind-the-scenes superhero–quiet, reliable, and absolutely essential."

How It All Comes Together

Azhar scribbled furiously in his notebook. "Okay, so we've got Calculus predicting the future, Trigonometry handling angles and paths, and Statistics organizing the data. But how do they all work together?"

"It's like a team," Raj said. "Think of Trigonometry as the scout–it measures distances and angles. Statistics is the strategist, organizing all the information we've got. And Calculus? That's the action hero, taking all this data and turning it into answers about rates of change, areas, and totals."

Azhar grinned. "So, basically, Trigonometry, Statistics, and Calculus are like the Avengers of math?"

"Exactly!" Raj said, laughing. "They team up to simplify complex problems, whether it's programming robots, forecasting the weather, or even optimizing how much pizza you can fit in a box."

"Wait, wait," Azhar interrupted, raising a hand. "Did you just say pizza? Now this is a math lesson I can get behind. Tell me Calculus can optimize how many toppings I can pile on my slice."

Raj smirked. "Not just toppings, my friend. Calculus can even maximize the volume of your pizza box so you can carry more slices!"

Azhar threw up his hands in mock despair. "Why didn't they teach us this in school? I would've been the Einstein of Pizza Physics by now!"

The Journey Ahead

As their conversation continued, Azhar's curiosity deepened. What had started as a fun discussion about robots and movies had turned into a realization: Math wasn't just about numbers on a board–it was the language of reality itself.

"So, what's next?" Azhar asked, standing up to leave.

Raj smiled. "Next, we learn how to use this math ourselves. Who knows? Maybe one day we'll build the next *Terminator* robot—or save the planet from global warming."

"Or," Azhar added with a mischievous grin, "we could just use it to invent the ultimate pizza box."

They both laughed, walking out of the café with a shared sense of excitement about the endless possibilities waiting for them in the world of math—and beyond.

How Calculus Can Help Predict Climate Change and Natural Disasters

Calculus, often described as the mathematics of change, plays a vital role in understanding and predicting the complex systems that govern our planet. With natural disasters like forest fires, floods, and rising global temperatures becoming increasingly common, calculus provides the tools needed to analyse and mitigate these challenges.

Modelling and Predicting Climate Change

1. Understanding Rates of Change:

- o Differential calculus is used to calculate rates of change, such as the speed at which temperatures are rising or ice caps are melting.

- For example, by analysing how CO2 levels increase over time, scientists can predict future atmospheric concentrations and their impact on global temperatures.

2. Tracking Accumulation:

- Integral calculus helps measure cumulative effects, such as the total amount of heat trapped in Earth's atmosphere over decades or the rise in sea levels caused by melting glaciers.

3. Analysing Weather Patterns:

- Calculus-based models simulate weather systems by solving differential equations that describe how heat, wind, and moisture interact in the atmosphere.

- These models help forecast extreme weather events, enabling governments to prepare for floods, hurricanes, or droughts.

Addressing Natural Disasters

1. Forest Fires:

- Using data on temperature, humidity, and wind speeds, calculus models predict the spread of wildfires. By understanding how fires propagate, firefighters can strategize containment measures.

2. Flood Predictions:

- o Calculus is used to model water flow in rivers and predict flooding patterns based on rainfall data. This allows for early warnings and better flood management strategies.

3. Long-Term Planning:

- o Predictive models powered by calculus help governments and organizations anticipate long-term impacts, such as water scarcity or agricultural losses due to changing climate conditions.

Inspiring Solutions

By understanding how changes occur in complex systems, calculus not only helps us predict future disasters but also inspires innovative solutions. For instance:

- **Renewable Energy:** Calculus optimizes solar panel efficiency and wind turbine placements to meet energy needs sustainably.

- **Global Warming Mitigation:** Models powered by calculus help identify the most effective strategies for reducing carbon emissions and slowing climate change.

Calculus is more than just math–it's a critical tool for safeguarding the future of our planet. By enabling us to understand and predict change with precision, it empowers humanity to confront its greatest challenges with science-driven solutions.

Why Learn Calculus?

Calculus is much more than a series of equations or a subject in a textbook. It is the language of change, a mathematical tool that helps us decode the dynamic and ever-evolving nature of the world around us. From understanding how planets orbit the sun to optimizing the design of the cities of tomorrow, Calculus teaches us to think critically, solve problems, and bridge the gap between abstract ideas and tangible realities.

A Toolkit for Innovation

Every field of human endeavour–from science to engineering, from medicine to economics–relies on the principles of Calculus to solve its most pressing problems. For example:

- **In medicine,** Calculus models the spread of diseases and optimizes drug dosages.
- **In business,** it forecasts market trends, helping companies maximize profits and minimize losses.
- **In engineering,** it designs safer, stronger structures and systems, from skyscrapers to bridges.

- **In space exploration,** it calculates rocket trajectories and ensures safe landings on distant planets.

By learning Calculus, you gain access to a universal toolkit that enables you to make sense of complexity, find patterns in chaos, and uncover solutions to the challenges that shape our future.

Unlocking the Secrets of the Universe

Calculus has been at the heart of humanity's greatest discoveries:

- It helped physicists like Newton and Einstein unravel the laws of motion and relativity.

- It is the foundation of technologies like artificial intelligence, robotics, and renewable energy.

- It continues to shape our understanding of the universe, from modelling black holes to predicting the fate of our climate.

Every curve on a graph, every shift in temperature, every heartbeat is a story of change that Calculus helps us understand.

Why It Matters

Learning Calculus is not just about solving mathematical problems. It's about developing a **mindset**–a way of thinking that embraces complexity and thrives on innovation. By mastering Calculus, you're learning to:

- Analyse situations with precision.
- Predict outcomes based on patterns and data.
- Think creatively to design solutions to real-world problems.

A Call to Action

Whether you're a budding scientist, an aspiring entrepreneur, or someone who simply wants to understand how the world works, Calculus is your gateway to unlocking potential.

By mastering the art of Calculus, you're not just solving equations—you're learning to **decode the mechanics of life, innovate for the future, and see the world in a way few others can.**

So, embrace the journey. With Calculus, you're not just studying math—you're preparing to **change the world.**

Statistics:

The Power of data and representation
"In God We Trust, For Everything Else We Need Data"

A Late-Night Revelation

Siddhant, the proud owner of a renowned fast-food chain, stood on the grand stage under the dazzling glow of the spotlight. The audience roared with applause, the sound of clapping echoing through the grand hall. Cameras flashed incessantly, capturing his moment of glory as he held the prestigious award aloft–a symbol of his culinary excellence and unmatched customer service.

Yet, as the cheers surged around him, Siddhant's smile faltered for the briefest moment, a crack in the armour that no one but himself could feel. The weight of the award in his hand seemed lighter than the heavy knot twisting in his chest.

He had worked tirelessly to build his empire from the ground up. Every recipe on the menu carried a part of his soul, and every smiling customer was a testament to his dedication. But deep down, beneath the glittering façade of success, he couldn't ignore the harsh truth gnawing at his conscience: sales were stagnating, and profits had begun to dwindle. The numbers, the lifeblood of his business, didn't align with the applause that surrounded him tonight.

As he descended the stage, the cheers fading into the background, Siddhant felt a hollow ache—a sense of dissonance between the world's recognition and his internal fears. It wasn't failure, but it wasn't victory either. Something was missing, something he couldn't quite name.

Hours later, in the hushed stillness of the hotel lobby, Siddhant sat alone, nursing a cup of lukewarm coffee. The grandeur of the event felt distant now, replaced by the stark reality of his worries. It was then that his old friend Ajay found him, slouched in an armchair, staring blankly at the city lights outside.

"Siddhant?" Ajay's voice was soft, yet filled with concern. "You should be celebrating tonight. What's wrong?"

Siddhant sighed, his voice heavy with unspoken frustration. "Ajay, everyone sees the award, but they don't see the struggle. My outlets are running, the customers seem happy, but the profits—they're slipping through my fingers. I don't understand where I'm going wrong."

Ajay studied his friend for a moment, then leaned forward with a knowing smile. "Siddhant, you have everything—great recipes, loyal customers—but have you looked at your data?"

"My data?" Siddhant frowned. "What do you mean?"

"The numbers don't lie, my friend. Sales trends, customer behaviour, menu preferences, complaints, demographics—it's all there, waiting to tell you the story you need to hear. The question is, are you listening?"

Siddhant's eyes widened slightly as Ajay's words began to sink in. "You think the answers are in the data?"

Ajay nodded. "Not just answers, Siddhant. Insights. Patterns. Solutions. The key to turning things around is already in your hands—you just need someone to help you decode it."

For the first time that evening, a flicker of hope crossed Siddhant's face. The road ahead wasn't clear yet, but Ajay's words had sparked a light. By the time he returned to his office the next day, he was no longer weighed down by doubt. Instead, he was armed with determination and a plan to uncover the hidden potential of his business.

That conversation had planted a seed of curiosity in Siddhant's mind. He hired a data analyst and began collecting, organizing, and analysing every bit of information from his outlets. What followed was nothing short of magical.

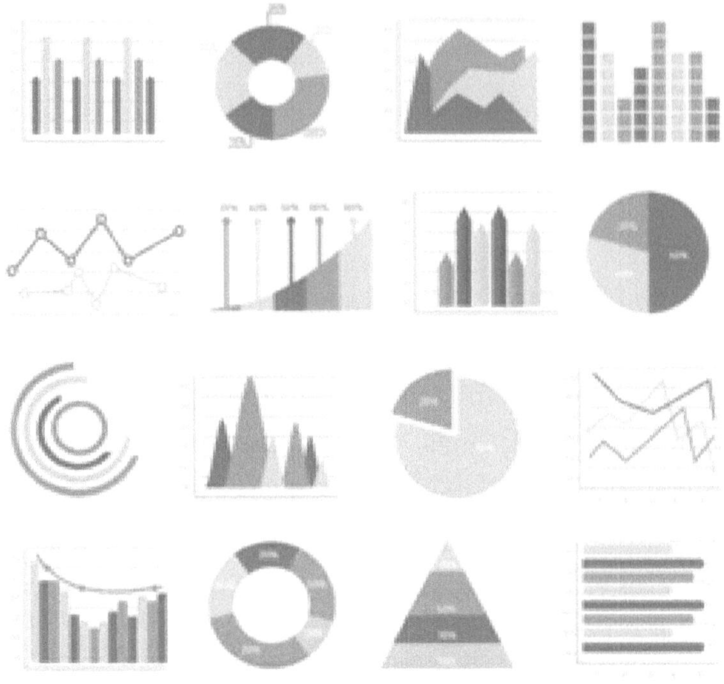

Picture credit: freepik.com

The Power of Data: Turning Insights into Action

Over the next few weeks, Siddhant's data analyst used advanced statistical methods to uncover hidden patterns and insights. Here's what they discovered and how it changed Siddhant's business:

1. Multivariate Analysis: This technique allowed Siddhant to analyse multiple variables simultaneously, such as:

- **Sales per outlet vs. footfall vs. demographics.**
- **Customer complaints vs. specific menu items vs. time of day.**

With this, he identified that outlets near colleges had high footfall but lower spend per customer. Meanwhile, locations in affluent areas had fewer customers but higher spending. This insight helped him tweak his menu and pricing strategies to maximize revenue.

2. Regression Analysis: Using regression analysis, Siddhant identified relationships between different variables, such as:

- How pricing changes influenced repeat customers.
- The impact of advertising spends on footfall.

Formula for regression:

$$Y = a + bX + \epsilon$$

Here, Y is the dependent variable (e.g., sales), X is the independent variable (e.g., advertising spend), a is the intercept, b is the slope, and ϵ is the error term.

Siddhant found that promotions on certain days led to higher weekend sales, so he shifted marketing efforts accordingly.

3. Correlation: The team used correlation to study the strength of relationships between variables. For instance:

- Strong correlation (r=0.85) between complaints about wait times and customer retention rates.
- Weak correlation (r=0.2) between new menu items and increased footfall.

This revealed that fixing service speed was more critical than experimenting with new recipes.

4. Deviation and Variance: Analysing deviation and variance helped Siddhant identify inconsistencies:

- Outlets with high variance in sales indicated unstable performance.
- Low deviation in customer satisfaction scores highlighted consistently good service.

Formula for Variance:

$$\text{Variance} = \frac{\sum(X - \bar{X})^2}{N}$$

This led him to implement standardized training programs across all outlets.

The Transformation

Armed with these insights, Siddhant made strategic changes:

- **Menu Optimization:** Reduced low-performing items and introduced combo deals at college locations.
- **Operational Efficiency:** Implemented faster service protocols in high-complaint outlets.
- **Targeted Marketing:** Focused ads on high-potential demographics.

Within months, Siddhant's fast-food chain had transformed into a data-driven powerhouse. By leveraging statistical analysis, he could pinpoint key factors such as customer preferences, peak hours, and inventory turnover. This enabled precise decisions about menu pricing, promotional strategies, and supply chain optimization. The insights derived from the data led to improved operational efficiency, smarter marketing campaigns, and better resource allocation. As a result, sales surged, customer satisfaction soared, and profits followed suit, making his business a resounding success. Data-driven decisions became his secret weapon, propelling the chain to victory in the competitive market.

Importance of Statistics in Everyday Life

When Mukesh Ambani famously declared that "data is the new oil," he referred to the immense value of data and bandwidth in the telecom industry, emphasizing its potential to fuel modern digital economies.

However, if we extend this metaphor to the field of statistics, the analogy becomes even more insightful.

Data in its raw form–like crude oil–is abundant but unrefined, often chaotic and unstructured. It is through the lens of statistical tools and techniques that this crude data is processed, cleaned, and analysed, transforming it into "usable oil" or **intelligent data**. Just as crude oil must go through refining processes to extract fuels like petrol or diesel, raw data requires statistical refinement–such as analysis, correlation, regression, and visualization–to yield actionable insights.

This refined, intelligent data becomes a powerful resource for decision-making, strategy development, and innovation across industries. Whether predicting market trends, optimizing business operations, or even combating global challenges like pandemics, statistics serves as the refinery that unlocks the true potential of raw data, making it as indispensable in modern problem-solving as oil is in powering our lives.

Statistics, often regarded as the language of data, is critical in making informed decisions. From businesses analysing customer behaviour to governments planning resources, statistics play a pivotal role. Some applications include:

Healthcare: The Detective That Never Sleeps:

Imagine a world without statistics in healthcare–it would be like trying to solve a mystery without a detective. When a sudden flu outbreak sweeps through a city, healthcare statisticians are like Sherlock Holmes, combing through mountains of data: patient histories, geographic hotspots, and the speed of transmission.

Take the case of COVID-19. It wasn't just masks and vaccines that helped us–it was data. Statisticians tracked infection curves, estimated R-values (how contagious the virus was), and even predicted when your neighbourhood might run out of toilet paper! They're the unsung heroes who ensure that healthcare isn't just about treating the sick, but about outsmarting diseases before they strike. Without them, we'd still be stuck diagnosing diseases with tea leaves and crystal balls.

Economics: Predicting the Unpredictable:

Economics without statistics is like trying to forecast the weather using a magic eight ball. Want to know if the market will crash next year? Or if consumers will suddenly stop buying avocado toast? Economists use statistical models to analyse trends and behaviours, crunching numbers like fortune-tellers with Excel sheets instead of crystal balls.

Take inflation, for example. Statistics tells us if prices are rising too fast, if people are spending too little, or why your coffee suddenly costs as much as a small car. And when it comes to consumer behaviour, statisticians dig into patterns that help businesses predict your next impulse buy. Yes, they know you'll probably buy that "unnecessary but oddly satisfying gadget" during a sale, because the numbers never lie.

Education: A Grade-A Problem Solver:

Picture a classroom without any way to measure student performance. Teachers would be grading homework based on whether the handwriting was neat or if the essay made them laugh. Enter statistics: the superhero of education. It helps track student progress, identify struggling learners, and allocate resources more effectively.

Imagine a school principal holding a colourful bar chart showing test scores–low in math, high in storytelling. They immediately decide to bring in a new math tutor while encouraging the budding Shakespeare in their midst. Statistics ensures that every student gets the help they need, even if they secretly believe that algebra is just a conspiracy to make their lives miserable.

Cricket: Where Legends Meet Numbers

Cricket isn't just a game of skill and passion–it's a battlefield of numbers, and the statistics behind the legends tell stories as gripping as the matches themselves. Let's dive into a hypothetical clash of cricket titans and see how statistics play a role in decoding their greatness.

Sachin Tendulkar vs. Brian Lara: The Battle of the Bats

Sachin Tendulkar, the "Little Master," and Brian Lara, the "Prince of Trinidad," were cricket's ultimate run machines. Fans still debate: who was better? While emotions often cloud judgment, statistics come to the rescue.

- **Average and Consistency:** Tendulkar's career batting average of 53.78 in Tests and his unmatched 100 international centuries make him the epitome of consistency. But Lara's highest Test score of 400* stands as a monument to his ability to single-handedly dominate an innings.

- **Match Impact:** Lara often played for a struggling West Indies team, and statistics reveal that a higher percentage of his centuries came in matches his team was losing–a testament to his ability to perform under pressure. Meanwhile, Tendulkar's centuries often came in tightly contested matches, making him a reliable anchor for India.

- **Fun Stat:** Tendulkar faced over 30,000 deliveries in his international career. That's like batting for nearly two straight years if you played eight hours a day, every day. Lara, on the other hand, seemed to save his best for special occasions–his career can be summed up as quality over quantity.

Kapil Dev: The All-Rounder vs. The Spinners' Duel

Switching gears to bowlers, Kapil Dev, India's "Haryana Hurricane," might smile slyly watching a hypothetical battle between Shane Warne and Muttiah Muralitharan, two of the greatest spinners ever.

- **Warne vs. Muralitharan:** Warne, with 708 Test wickets, relied on tactical brilliance, flight, and turn, often deceiving batsmen into playing false shots. Muralitharan, with his record 800 Test wickets, was the master of relentless accuracy, turning pitches into a spinner's paradise.

- **Strike Rates:** Warne struck every 57 balls in Tests, while Muralitharan's strike rate of 55 balls showed his ability to clean up batsmen just a touch faster. In ODI cricket, however, Muralitharan's economy rate of 3.93 kept batsmen guessing far more effectively than Warne's 4.25.

- **Head-to-Head Drama:** Imagine Tendulkar facing Warne, who famously struggled to dismiss him (Tendulkar had a batting average of over 60 against Warne). Now, switch to Muralitharan bowling to Lara, who once scored a blistering 153 against Sri Lanka under Muralitharan's watch—proof that even the best bowlers are at the mercy of genius batsmen.

Kapil Dev's Place in the Mix

Kapil Dev stands apart as an all-rounder, contributing with both bat and ball. His 434 Test wickets paired with 5,248 Test runs make him a rare gem. Statistics show his ability to win matches single-handedly:

- His unbeaten 175 in the 1983 World Cup against Zimbabwe is still celebrated as one of the greatest ODI innings of all time.

- Bowling stats reveal that his knack for breaking partnerships often came in the most crucial moments.

Drama Through Numbers

Imagine a fantasy match where:

- Tendulkar's sublime technique meets Warne's crafty leg-spin.
- Lara's attacking flair takes on Muralitharan's unerring accuracy.
- Kapil Dev, tasked with balancing the team, charges in with the new ball or steadies the ship with the bat.

Statistics wouldn't just tell you who played better—they'd decode strategies:

- What lengths Warne should bowl to Tendulkar to avoid boundaries.
- How Muralitharan's variations could trap Lara.
- How Kapil Dev's batting strike rate could rescue his team during a collapse.

Conclusion: Numbers Beyond the Game

Statistics don't just measure greatness—they provide context. They show why Warne was so tactical, why Muralitharan was so effective, and why Tendulkar and Lara were cricket's eternal rivals in excellence. Cricket may be a game of passion, but in the end, it's the numbers that decide who walks off the field as the true champion.

Applications of Statistical Tools

Statistical tools help us make sense of data, find patterns, and make informed decisions. Here's an in-depth look at the key categories with examples and tools:

1. Descriptive Statistics: Summarizing Data

What It Does:

Descriptive statistics provides a snapshot of data by summarizing it through key measures like mean, median, mode, range, and standard deviation. This helps in understanding the distribution and spread of data.

Key Measures:

- **Mean (Average):** Adds up all values and divides by the count. Useful for finding the average score, sales, etc.

- **Median:** The middle value when data is sorted. Helps when data has outliers (e.g., income distribution).

- **Standard Deviation:** Measures data spread around the mean. A low value indicates data points are close to the mean, while a high value suggests wide dispersion.

Tools:

- **Excel:** Use functions like =AVERAGE,=MEDIAN, and =STDEV to compute these.

- **Python**: Libraries like NumPy and Pandas are perfect for calculating descriptive statistics.

Example: A restaurant tracks daily footfall for a month:

Example: A restaurant tracks daily footfall for a month:

- Footfall Data: 100, 120, 130, 150, 140, 180, 170...
- Mean Footfall: $\frac{\text{Sum of Footfalls}}{\text{Number of Days}} = 140$
- Median Footfall: Middle value (e.g., 140 when sorted).
- Standard Deviation: If $\sigma = 15$, most days fall within 140 ± 15 (125 to 155).

2. Inferential Statistics: Drawing Conclusions

What It Does:

Inferential statistics takes sample data and generalizes or predicts about the larger population. It involves:

- **Hypothesis Testing:** Testing assumptions using statistical methods.

- **Confidence Intervals:** Providing a range of values within which a population parameter is likely to fall.

- **Significance Testing:** Determines if observed differences are statistically significant.

Tools:

- **SPSS (Statistical Package for the Social Sciences):** Great for hypothesis testing and regression.

- **R:** Widely used for conducting inferential analyses and visualizing results.

Example: Imagine a start-up surveying 500 customers to predict behaviour across 50,000 customers:

- **Hypothesis:** Customers prefer online ordering over in-store purchases.

- **Survey Result:** 70% prefer online ordering.

- **Confidence Interval:** With 95% confidence, the preference rate is between 68% and 72%.

- **Decision:** Based on this analysis, the start-up invests in its online ordering platform.

3. Predictive Analytics: Forecasting Future Outcomes

What It Does:

Predictive analytics uses advanced techniques like regression, correlation, and machine learning to forecast trends and behaviours.

Key Methods:

- **Regression Analysis:** Predicts relationships between variables (e.g., how advertising spend impacts sales).

- **Formula:** $Y = a + bX + \epsilon$

Here, Y is the dependent variable (e.g., sales), X is the independent variable (e.g., advertising spend), a is the intercept, b is the slope, and ϵ is the error term.

- **Correlation**: Measures how strongly two variables are related (value ranges from -1 to +1).

Tools:

- **Tableau:** For visual analytics and trend forecasting.

- **Python/Scikit-learn:** For machine learning-based predictive modelling.

Example: A fast-food chain uses predictive analytics to optimize menu pricing:

- **Regression:** Predicts how price changes affect sales volume.
    - o **Data:** X = Price, Y = Sales.
    - o **Analysis:** Increasing burger prices by 10% results in a 5% drop in sales, but overall revenue increases.
- **Correlation:** Strong positive correlation (r=0.9) between advertising spend and footfall.
- **Decision:** Allocate more budget to advertisements on weekends for higher footfall.

Conclusion:

Statistical tools are indispensable across industries. Whether summarizing current data (descriptive), making generalizations (inferential), or predicting future trends (predictive), these methods empower decision-makers to act confidently and strategically. Tools like Excel, Python, and Tableau make these processes efficient and scalable, opening up possibilities for data-driven success.

Career Opportunities in Statistics: Unlocking the Power of Data

The rapid growth of data-driven decision-making has made careers in statistics not just lucrative but also incredibly exciting. Whether you're interpreting trends, building predictive models, or designing cutting-edge algorithms, the opportunities are as diverse as the industries they serve. Let's explore these roles in greater detail:

1. Data Analysts: The Storytellers of Data

What They Do:

Data analysts are the bridge between raw data and actionable insights. They sift through datasets, clean and organize them, and uncover patterns that help organizations make smarter decisions.

Exciting Opportunities:

- **Retail:** A data analyst at an e-commerce giant like Amazon might analyse customer purchase trends, discovering that certain products sell more during specific seasons or in particular regions. Based on this, they recommend inventory adjustments to avoid stockouts or overstocking.

- **Sports**: Analysing player performance metrics to advise coaches on strategies. For example, determining why a cricket team wins more matches when their star batsman scores above 50 runs in a game.

Tools Used:

Excel, SQL, Python, and visualization tools like Tableau or Power BI.

2. Statisticians: The Architects of Experiments

What They Do:

Statisticians are the backbone of data science, designing experiments, sampling data, and creating models to test hypotheses. They dig deep into data, ensuring every analysis is rooted in mathematical rigor.

Exciting Opportunities:

- **Pharmaceuticals:** Statisticians in companies like Pfizer or Moderna play a critical role in clinical trials. They analyse the effectiveness of new vaccines, ensuring that the results are statistically significant and reliable before launch.

- **Public Policy:** Working for governments to study census data and forecast resource needs like healthcare and education facilities over the next decade.

Tools Used:

R, SAS, SPSS, and advanced probability models.

3. Machine Learning Engineers: The Pioneers of AI

What They Do:

Machine learning engineers use statistics, programming, and domain knowledge to create algorithms that can learn and make predictions. These professionals are shaping the future with intelligent systems that drive automation and innovation.

Exciting Opportunities:

- **Self-Driving Cars:** Imagine working for Tesla, designing algorithms that teach cars to recognize traffic signs and avoid pedestrians–statistics in action, saving lives.

- **E-commerce Recommendations:** Building recommendation systems that suggest products based on past purchases. Ever wondered how Netflix knows you'll love that new series? It's a machine learning engineer at work.

Tools Used:

Python, Scikit-learn, TensorFlow, and cloud platforms like AWS or Google Cloud.

4. Business Intelligence Analysts: The Decision-Making Wizards

What They Do:

BI analysts focus on turning data into dashboards and reports that executives can use to make strategic decisions. They connect the dots between past performance and future goals.

Exciting Opportunities:

- **Banking:** Helping financial institutions like Goldman Sachs identify risks by analysing credit data and predicting defaults.
- **Hospitality:** Advising hotel chains like Marriott on pricing strategies by analysing occupancy rates and market demand trends.

Tools Used:

Tableau, Power BI, SQL, and data warehousing tools like Snowflake.

Beyond Traditional Roles: Specialized Opportunities

- **Data Ethics Specialists:** Ensuring ethical use of data in AI and analytics, especially in sensitive fields like healthcare or law enforcement.

- **Geospatial Analysts:** Using spatial data to advise on urban planning or disaster management, such as predicting flood-prone areas.

- **Social Media Analysts:** Measuring the effectiveness of marketing campaigns and predicting trends by analysing user interactions on platforms like Twitter and Instagram.

Why It's Exciting

- **Impact:** Your work can drive business decisions, influence public policy, or even save lives in healthcare and disaster management.

- **Creativity Meets Logic:** Combine analytical thinking with storytelling, creating solutions that are both impactful and innovative.

- **High Demand and Pay:** Statistics professionals are in demand across industries, offering excellent career growth and competitive salaries.

Conclusion

The growing importance of data in today's world means that careers in statistics are not just about crunching numbers–they're about solving real-world problems. Whether it's helping businesses thrive, saving lives in healthcare, or building futuristic AI systems, statisticians, analysts, and engineers are at the forefront of progress. With endless opportunities and exciting challenges, a career in statistics promises to be as rewarding as it is impactful.

******

A further insight for aspiring individuals wanting to make a career in this field:

National Informatics Centre (NIC): A Gateway to Digital India

The **National Informatics Centre (NIC)**, under the Ministry of Electronics and Information Technology (MeitY), is the technology backbone of the Indian government. Established in 1976, NIC has been pivotal in driving India's digital transformation by offering IT support, e-governance solutions, and data management services to government departments and institutions.

Key Areas of Work

NIC plays a critical role in:

1. **E-Governance:** Designing and implementing digital platforms like eHospital, eCourts, and the National Knowledge Network.

2. **Infrastructure Management:** Hosting and managing government websites and cloud services through its National Cloud - MeghRaj.

3. **Cybersecurity:** Ensuring the safety of government data with advanced cybersecurity measures.

4. **Data Analytics:** Leveraging big data and AI to inform policy decisions and improve governance.

5. **Emerging Technologies:** Exploring blockchain, IoT, and AI for government applications.

Career Opportunities at NIC

Working at NIC offers professionals the chance to contribute to national progress through technology. Some exciting roles include:

1. **Scientific Officer/Engineer:** Developing software and managing IT projects.

2. **Data Scientist:** Analyzing large datasets to provide actionable insights for government policies.

3. **Cybersecurity Specialist:** Protecting critical government infrastructure from threats.

4. **Network Engineer:** Managing and optimizing India's vast governmental IT network.

5. **AI/ML Specialist:** Creating AI-driven solutions for governance challenges.

Why Join NIC?

- **Impactful Work**: Contribute to projects that directly improve citizens' lives.

- **Skill Development:** Work with cutting-edge technologies like AI, IoT, and cloud computing.

- **Job Stability:** A government organization offering secure, long-term career prospects.

- **Competitive Pay:** NIC provides attractive remuneration packages as per central government norms.

How to Join NIC

Recruitment is typically conducted through competitive exams by the **National Institute of Electronics and Information Technology (NIELIT)** or direct hiring for specific projects. Aspiring candidates should keep an eye on the official website (www.nic.in) for updates on job notifications.

A career at NIC is not just a job—it's an opportunity to contribute to India's digital future while working on transformative projects that make a real difference.

The Four dimensions and beyond:

A glimpse into the future

Aryan and Salil sank into the plush chairs of their favourite café, the soft hum of chatter mingling with the aroma of freshly brewed coffee. Sunlight streamed in through the window, casting a golden glow over their table as Aryan leaned forward, his eyes sparkling with curiosity. He cradled his coffee cup like a prop in a suspenseful monologue.

"Do you remember *Back to the Future?* The late 80s? Marty McFly, the DeLorean, *time travel?*" Aryan's voice carried a sense of wonder, as though he were unravelling a great mystery.

Salil grinned wide, stirring his coffee with a dramatic flourish. "How could I forget? Marty zooming through time, trying to fix his past, save his future... wild stuff back then. But now?" He leaned back with an exaggerated pause. "It doesn't seem so impossible anymore, does it?"

Aryan's smile widened. "Exactly, Salil. Back then, it was just science fiction—wild dreams on a Hollywood screen. But think about it! Time isn't just a plot device anymore. It's the *fourth dimension*—something real, something we can *measure, bend,* and maybe someday... even *control.*" His voice dropped to a whisper, as if sharing a secret too big for the room to handle.

Salil sat up, suddenly hooked. "Control time? Like *the Matrix* meets Einstein?"

Aryan nodded, his enthusiasm spilling over. "Yes! Time, Salil—it's everywhere. It governs us like a silent ruler. Just like length, breadth, and height define the world we see, time gives us motion, change, life. Without it, we'd be frozen. But what if we could manipulate it? What if AI, quantum physics, and technology are already nudging us in that direction? You've seen Musk's rockets land back *on Earth*, right? That's not magic—it's math. It's time *calculated* to perfection."

Salil's coffee sat untouched as he absorbed Aryan's words. "So you're saying... the stuff of movies is becoming real?"

Aryan's eyes burned with intensity now. "It already is. Without math, we wouldn't launch rockets, design self-driving Teslas, or simulate alternate realities like *The Matrix*. Time and space are no longer mysteries—they're tools. And we're learning to wield them."

He pointed out the window, where a plane soared across the sky. "That's what the future looks like, Salil—a world where time, like any other dimension, becomes something we *understand, measure,* and *master*. It won't just be science fiction anymore."

Salil whistled low, shaking his head with awe. "Man, I thought I was here for a coffee. You're serving up some serious sci-fi future."

Aryan grinned. "It's not sci-fi, my friend. It's science—mathematics. And the future? Well, it's just waiting for us to catch up."

The two friends sat there, their imaginations spinning faster than the Earth's rotation, caught in the thrilling realization that the boundaries of space and time were no longer limits—they were possibilities.

Dimensions – The Fabric of Our Reality

Salil raised an eyebrow, "Alright, professor Aryan, you're going deep again. Explain to me—why are dimensions such a big deal?"

Aryan chuckled. "Think about it. The three spatial dimensions—length, breadth, and height—form the foundation of everything around us. Every building, car, or phone you use is designed using these dimensions. Without mathematics, how would engineers at Tesla build cars or design rockets that land back on Earth?"

"True," Salil said, nodding. "Elon Musk didn't create SpaceX and Tesla with sheer imagination. He used math—calculus, geometry, and physics—to manipulate these dimensions in real life."

Aryan continued, "And then there's time—the fourth dimension. It changes everything. In *Back to the Future,* time travel was fiction, but look at today: AI and quantum physics are helping scientists calculate scenarios where time *could* behave differently. Even Einstein showed us that time and space are interconnected. Without mathematics, his theory of relativity would just be scribbles on paper."

The Matrix, AI, and Quantum Physics

"Okay, so where does *The Matrix* come in?" Salil asked, curious now.

Aryan leaned back, "Ah, *The Matrix*. It's not just about simulations and alternate realities. The movie asked us to rethink what reality even is. Today, we use AI and quantum physics to build models that simulate the real world. For example, AI can predict weather patterns or help Tesla's self-driving cars see dimensions. Every pixel on a computer screen or in virtual reality is manipulated using mathematics—linear algebra, matrices, and geometry."

Salil was intrigued. "So, quantum physics and math can help us predict the future?"

"Kind of," Aryan replied. "Quantum physics studies particles at a subatomic level, where time and space behave strangely. It's the key to future technologies like quantum computers. Google and IBM are already building these machines, which will process data exponentially faster than current computers. And the math driving quantum physics—vectors, matrices, probabilities—will shape the future."

Picture credit: freepik.com

A Future Written in Dimensions

"So, what's next? Where are the four dimensions taking us?" Salil asked, his voice carrying both wonder and scepticism.

Aryan smiled. "The future is already here, Salil. Let me paint you a picture:"

1. Space Exploration: Navigating the Final Frontier

Space exploration is a masterclass in four-dimensional mathematics—length, breadth, height, and, most critically, time. SpaceX, NASA, and emerging private players like Blue Origin are redefining humanity's reach into the cosmos.

- **Trajectories in Four Dimensions:** Calculating the perfect launch window is a mind-boggling feat of math. Every millisecond matters—engineers predict when a rocket must leave Earth, where it will intersect with Mars's orbit, and how long it will take to land safely. This requires solving thousands of equations to account for Earth's rotation, gravitational pulls, and planetary positions.

- **Example:** The Mars Rover Perseverance didn't just "land" on Mars—it touched down after 7 months of flight, where every move was planned using physics, calculus, and time-based predictions.

- **The Future:** NASA's Artemis mission aims to return humans to the Moon and beyond, while SpaceX's Starship is being developed to make interplanetary colonization a reality. Without precise mathematics governing space-time, we wouldn't even leave the launchpad.

2. AI and Robotics: Machines That 'See' and 'Think'

Artificial Intelligence (AI) and robotics function in multi-dimensional spaces, and mathematics is the fuel that drives their decision-making capabilities.

- **Multi-Dimensional Learning:** AI algorithms rely on linear algebra (vectors, matrices) to process data and make decisions. Robots "see" the world by analysing millions of data points in real-time, predicting movements and understanding environments.

- **Example:** Self-driving cars like those made by Tesla analyse their surroundings in real time. They map the 3D world using LIDAR (Light Detection and Ranging), sensors, and cameras, while time determines their decisions: when to brake, when to swerve, and when to accelerate. Without advanced mathematics, these decisions would be guesswork.

- **Humanoid Robots:** Boston Dynamics' robots, like Atlas, use differential calculus and motion dynamics to balance, jump, and even dance. Their actions are planned in multi-dimensional space where every move depends on time and space synchronization.

3. Virtual and Augmented Reality: Creating Parallel Worlds

Virtual Reality (VR) and Augmented Reality (AR) are transforming how we interact with technology by simulating real-world experiences.

- **Mapping Reality:** VR recreates our three-dimensional world using geometry and linear algebra. Every object is built using 3D coordinates (length, breadth, and height), while time adds movement, ensuring experiences occur in real-time.

- **Example:**
    - o In gaming, titles like *Half-Life*: Alyx and immersive VR experiences rely on math to create realistic movements, shadows, and physics.

- AR tools like Google Maps' Live View overlay digital elements onto the real world in real time, enhancing navigation.

- **The Future:** The Metaverse, a concept championed by companies like Meta (Facebook), will create a fully immersive digital world where 3D spaces interact seamlessly with time, allowing meetings, entertainment, and learning experiences to occur simultaneously across the globe.

4. Medicine and Healthcare: Transforming Lives with Precision

Mathematics and four-dimensional analysis have revolutionized medicine, offering life-saving tools for diagnosis, treatment, and prevention.

- **4D Imaging:** Traditional scans provide 2D or 3D images, but 4D imaging integrates time, showing how organs move, blood flows, and tumours grow. This helps doctors analyse critical changes in the body over time.

- **Example:**

    - Cardiologists use 4D echocardiography to observe the heart's real-time movements during surgery.

    - In cancer treatment, AI tools analyse thousands of patient records to predict tumor growth patterns over time and optimize therapies.

- **Robotic Surgeries:** Surgical robots like the Da Vinci system rely on precise mathematical models to perform surgeries, minimizing errors. They incorporate real-time data to adapt to tiny movements, enhancing accuracy.

- **The Future:** Predictive AI models will soon anticipate diseases before symptoms appear, transforming medicine from reactive to proactive. Imagine being warned of a potential heart attack weeks before it happens—math and AI make this possible.

5. Smart Cities: Living in the Cities of Tomorrow

Smart cities are the future of urban living, where technology, data, and mathematics converge to create efficient and sustainable environments.

- **Real-Time Optimization:** Time, as a dimension, is critical in managing energy, traffic, and resources. AI-powered systems analyze data in real time to make cities "think" and respond dynamically.

- **Example:**
    - Traffic Management: AI algorithms predict traffic patterns using real-time GPS data. Cities like Singapore have implemented smart traffic systems that adjust signal timings based on vehicle flow to reduce congestion.
    - Energy Efficiency: Smart grids monitor energy consumption and distribute electricity efficiently, preventing wastage during peak hours.

- **Urban Planning:** Architects and engineers design cities in 3D using tools like BIM (Building Information Modelling). Time (the fourth dimension) helps simulate how cities will evolve over years—accounting for growth, climate change, and infrastructure needs.

- **The Future:** Imagine a world where AI predicts accidents before they happen, public transport arrives precisely when needed, and energy grids balance themselves based on usage patterns—all powered by mathematics and real-time data.

The Role of Mathematics: The Silent Engine

Salil shook his head in awe. "And all of this depends on mathematics? Even Elon Musk's rockets and AI?"

Aryan nodded. "Absolutely! Mathematics is the silent engine that powers all progress. Without it, we wouldn't have the tools to measure length, analyse time, or manipulate space. From the construction of a simple house to the exploration of new galaxies, math is the language that makes everything possible."

He paused, then added with a grin, "The universe itself is a mathematical masterpiece. It's up to us to decode it."

Takeaway:

As the sun dipped below the horizon, Salil stared thoughtfully out the window. "So, dimensions aren't just for scientists or mathematicians. They're the foundation of the future."

Aryan smiled. "Exactly. The four dimensions—length, breadth, height, and time—define everything around us. They are no longer abstract concepts. They're tools to build, explore, and shape the world of tomorrow. With mathematics as our guide, there's no limit to what we can achieve."

"Maybe one day we *will* have our own DeLorean," Salil joked.

Aryan raised his coffee cup. "To the future, my friend—the one written in numbers and dimensions."

And as the two friends toasted, the quiet hum of possibilities filled the air—a future shaped by time, space, and the timeless power of mathematics.

The Future with Mathematics: Unlocking the Secrets of the Universe

As we stand on the precipice of the future, mathematics emerges as the ultimate key to unravelling mysteries that stretch beyond the boundaries of our imagination. From **dark matter** and **hyperspace** to the tantalizing possibilities of time travel, mathematics holds the power to decode the unknown and shape the future of humanity.

1. Dark Matter: Mapping the Invisible

Scientists estimate that **85% of the universe** is made up of **dark matter**—an invisible substance that doesn't emit light or energy but exerts gravitational pull on galaxies. Without mathematics, this cosmic enigma would remain beyond comprehension.

- **How Math Helps:** Physicists use complex models involving differential equations and gravitational theories to detect dark matter indirectly by studying how light bends (gravitational lensing) around unseen masses.

- **The Future:** Imagine a time when advanced mathematics and AI allow us to "map" dark matter, helping us understand how galaxies form, expand, and evolve. It could rewrite our understanding of the cosmos and reveal what truly holds the universe together.

2. Hyperspace: Beyond the Dimensions We Know

The idea of hyperspace–dimensions beyond our known three spatial dimensions–has fascinated scientists and dreamers alike. Mathematics is the bridge between reality and the possibility of these higher dimensions.

- **Current Exploration:** String theory, a ground-breaking area of physics, suggests that the universe operates in **10 or 11 dimensions**. These theories rely heavily on mathematical tools like **tensor calculus**, matrices, and differential geometry.

- **The Future:** Hyperspace could one day enable **interstellar travel.** If we master higher dimensions mathematically, humans might bypass the constraints of light-speed travel, creating "shortcuts" through the universe using **wormholes**–an idea born from Einstein's equations.

Imagine future spacecraft "folding" space to jump across galaxies in seconds, a concept once confined to science fiction but now grounded in mathematical possibilities.

3. Time Travel: Fact or Fiction?

The fourth dimension–time–has intrigued us for centuries. While time flows forward for us, mathematics suggests it might be possible to manipulate it.

- **Einstein's Legacy:** Albert Einstein's **General Theory of Relativity** showed that space and time are interconnected as **spacetime**. Under extreme gravitational forces, time behaves differently. For instance, astronauts in space experience time slower compared to those on Earth–a phenomenon called time dilation.

- **The Math Behind It:** Solutions to Einstein's equations suggest that **closed time-like curves** (loops in spacetime) could exist, theoretically enabling time travel. Scientists are exploring this with complex models using advanced geometry and quantum physics.

- **The Future:** While building a time machine still sounds far-fetched, mathematics and quantum theories may one day let us glimpse or even influence the past and the future. Imagine sending information backward in time to prevent disasters or solve global challenges.

4. Quantum Physics and the Multiverse

Quantum mechanics, the study of particles at subatomic levels, is rewriting reality as we know it. Mathematics is its language, helping us understand a world where particles exist in **multiple states simultaneously**–a phenomenon called superposition.

- **The Multiverse Hypothesis:** Some theories suggest that our universe is one of **infinite parallel universes**–each existing in a separate quantum state. Mathematical models like those in quantum field theory and probability distributions are exploring this mind-bending idea.

- **The Future:** Could we one day communicate with alternate universes? Mathematics might hold the key to proving their existence and unravelling the mysteries of life, reality, and consciousness.

5. AI and the Universe: A New Dawn

Artificial Intelligence is poised to become the ultimate mathematical tool, capable of solving equations and problems that are currently beyond human ability.

- **Predicting the Unknown:** AI-driven models, powered by advanced calculus and neural networks, can simulate entire galaxies, predict cosmic events like supernovae, and even map **black holes**.

- **Human Evolution:** AI may help us simulate complex systems like **human consciousness** and life itself, leading to discoveries about what truly makes us human.

Imagine AI working alongside quantum computers to solve the mysteries of spacetime, dark energy, and hyperspace. The possibilities are limitless.

The Grand Conclusion: A Universe of Infinite Potential

Mathematics is the silent force behind every breakthrough, from mapping the universe to understanding the very fabric of reality. It gives us the tools to decode dark matter, explore hyperspace, bend time, and perhaps even interact with parallel universes.

What does the future hold?

- A world where **space travel** is routine, powered by mathematical models.

- A time when **dark matter and dark energy** are understood, revealing the universe's hidden structure.

- A reality where **time travel** and alternate dimensions aren't science fiction but scientific achievements.

As Aryan might say to Salil over coffee:

"The universe is like a vast puzzle written in the language of mathematics. Each discovery is a piece of the puzzle, and with time, we'll unlock it all—one equation at a time."

In the end, mathematics isn't just a tool; it's our gateway to the infinite. The future doesn't wait–it's up to us to decode it.

Epilogue:

The Journey Beyond Numbers

As we come to the close of this book, it's important to remember that numbers, equations, and dimensions are more than abstract ideas—they are the unseen forces shaping the world we live in and the future we strive to build. From the smallest particles of quantum physics to the vast expanse of hyperspace, mathematics remains humanity's most powerful tool for understanding, innovating, and growing.

Throughout these pages, we explored how data drives businesses, dimensions shape our reality, and statistical tools unlock hidden patterns. We saw how technology, powered by math, is turning fiction into fact—whether it's predicting diseases, building smart cities, or propelling us toward the stars. The future is not just waiting to be discovered; it is waiting to be *calculated, created,* and *mastered*.

But beyond formulas and theories lies the essence of this book: the idea that knowledge has no boundaries. Each chapter reflects how mathematics transcends fields, disciplines, and even imagination, revealing possibilities that once seemed impossible. The stories within are not just about equations; they are about people—dreamers, explorers, and innovators who use mathematics to solve problems and build a better world.

As you close this book, let it be the beginning of your own journey. Whether you are a student, a professional, or simply someone curious about the universe, remember this: the world is written in the language of numbers, and each of us holds the power to decode it. The challenges of tomorrow—be it exploring new frontiers, solving global issues, or redefining reality—will not be conquered with guesswork but with insight, perseverance, and the timeless brilliance of mathematics.

In the end, as we navigate through the dimensions of time and space, one truth remains constant: the future belongs to those who can see the patterns, ask the right questions, and harness the power of knowledge.

And so, the journey doesn't end here—it only begins.

My magic formula –

Acknowledgement

My heartfelt gratitude and deepest appreciation go out to each and every one of you who stood by me like a rock through the turbulent storms and calm waters of my life. In moments of need, your unwavering support–whether financial, emotional, or just being a comforting presence–became the light that guided me through the darkest days.

You have been my winning formula, my secret equation to success. These relationships are the ingredients and elements of my mathematics, the constants in a life full of variables, and the solutions that simplified my journey. Each of you has played a unique role—sometimes as inspiration, other times as motivation—and always as a reminder that no matter how complex life's problems seemed, they could be solved with your love, guidance, and encouragement.

Your kindness and belief in me have been the fuel for my perseverance, lifting me up when I stumbled and reminding me that life's ups and downs are just part of the beautiful equation we solve every day. Thank you for being the balance, the strength, and the reason my journey has felt so meaningful. This chapter, and every success I celebrate, is shared with all of you.

Family:

- Bimla Devi Gupta (Mother)
- Lakhi Prasad Gupta (Father)
- Sanjana Gupta (Wife)
- Tanishka Gupta (Daughter)
- Khyati Gupta (Daughter)
- Anup Gupta (Brother)

Cousins:

- Naresh Kejriwal
- Anil Gupta
- Sanjay Puria

Friends:

- Vishal Gwalani
- Dr. Shailendra Pundale
- Sridhar Iyer
- Sanjay Parekh
- Kamlesh Rao
- Ajay Marwah
- Parag Dandekar
- Vikas Arora
- Aditya Bawa
- Rishi Deshpande

- Pravin Bhat
- Sameer Fegade
- Chet Ahuja
- Usha Dandekar
- Sandhya Shankar
- Niraj Gunde
- Anand Birje
- Milind Dighe
- Pawan Shukla

References& Resources:

1. **Book:** Fooled by Randomness : The Hidden Role of Chance in Life and in the Markets by Nassim Nicholas Taleb

2. **Book:** Built to Last by Jim Collins and Jerry Porras.

3. **https:**//www.sciencealert.com/how-a-genius-formula-carved-onto-a-bridge-changed-mathematical-history: The article discusses a remarkable mathematical formula carved onto the Broome Bridge, Dublin. This formula played a significant role in advancing mathematical understanding, particularly in relation to the development of calculus and mathematical analysis. It highlights how this inscription demonstrated the application of theoretical concepts in a practical context, influencing future mathematicians and contributing to the evolution of mathematics. The story emphasizes the intersection of art, engineering, and mathematics in shaping history and furthering scientific knowledge.

www.ingramcontent.com/pod-product-compliance
Ingram Content Group UK Ltd.
Pitfield, Milton Keynes, MK11 3LW, UK
UKHW040243300726
14061UKWH00002BD/133

9 789358 987263